Desert Anzacs:
The Under-Told Story of the Sinai Palestine Campaign, 1916-1918

Neil Dearberg

Glass House Books

Glass House Books
an imprint of IP (Interactive Publications Pty Ltd)
Treetop Studio • 9 Kuhler Court
Carindale, Queensland, Australia 4152
sales@ipoz.biz
ipoz.biz/ipstore

First published by IP in 2017
© Neil Dearberg, 2017

All rights reserved. Without limiting the rights under copyright reserved above, no part of this publication may be reproduced, stored in or introduced into a retrieval system, or transmitted, in any form or by any means (electronic, mechanical, photocopying, recording or otherwise), without the prior written permission of the copyright owner and the publisher of this book.

Printed in 11 pt Book Antiqua on 14 pt Avenir Book.

National Library of Australia
Cataloguing-in-Publication entry:

Creator:	Neil Dearberg – author
Title:	Desert Anzacs; the under-told story of the Sinai Palestine campaign, 1916-1918 / Neil Dearberg.
ISBN:	9781925231625 (paperback) 9781925231632 (eBook)
Notes:	Includes bibliographical references
Subjects:	Australia. Army. Australian Imperial Force (1914-1921) World War, 1914-1918 --Campaigns--Egypt--Sinai World War, 1914-1918--Campaigns--Palestine. World War, 1914-1918--Australia.

*To the unheard voices
To whom their nations owe so much.*

To Berin and Danelle from a proud Dad

*God and soldier, we adore, in time of danger, not before.
The danger passed and all things righted,
God is forgotten and the soldier slighted.*
— Rudyard Kipling

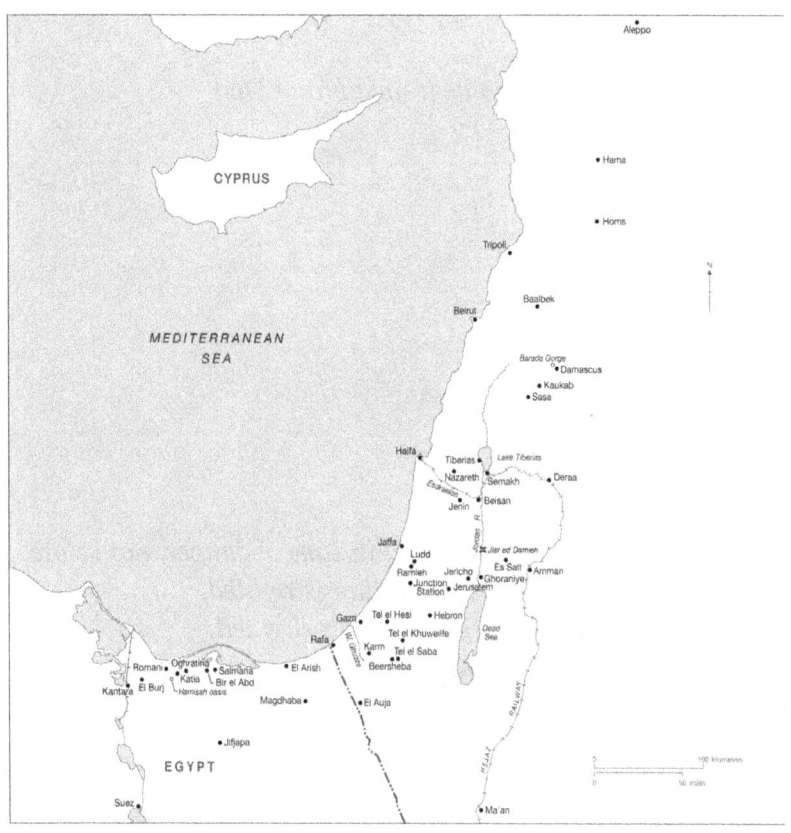

Figure 1: Theatre of Operations (map supplied by Department of Veterans' Affairs).

Contents

Author's Note	*vii*
Acknowledgements	*ix*
Abbreviations and Terms	*xi*
Introduction	*1*
Part One: From Home and into Sinai	**7**
1: Lose Suez…Lose the War	*9*
2: Desert Commander, Harry Chauvel	*14*
3: Getting to War	*17*
4: The Good, the Bad and the Arabs	*29*
5: The Anzac Journey	*41*
6: A New Chapter to the Anzac Legend	*54*
7: The Anzac Warriors	*69*
Part Two: Sinai to Gaza	**83**
8: Survival in Sinai	*85*
9: The Arab Revolt	*92*
10: Battle of Romani	*100*
11: Gaza: Two Tales of a City	*137*
12: The Arabs Have Aqaba	*156*
Part Three: A New Commander-In-Chief: General Allenby, June 1917 - October 1918	**169**
13: Gaza in Allenby's Sights	*171*
14: The Battle of Beersheba Opens Gaza	*179*
15: Gaza into Palestine	*190*
16: 'Raids' Across the Jordan	*211*
Part Four: Armaggedon	**229**
17: Who Wants Palestine?	*231*
18: Crush the Enemy	*246*

19: Plan, Prepare, Deceive	*261*
20: Let the Fight Begin	*269*
21: Their War is Over	*291*
Epilogue	*295*
Bibliography	*303*

Author's Note

People and place names are often spelt differently, depending on the nationality of an author or spelling preference. The terms 'Ottoman' and 'Turk' are interchanged in many texts, as they are here. Distances have been listed in various formats; feet, yards and metres depending on the source material.

To make this book more readable, I have avoided military and political jargon and technical details where possible. It is therefore less military and more personal than other texts about war and the stories of soldiers.

In the past, many historians and authors have been careful not to criticise British political and military leaderships. I offer criticism where it is warranted and I do not shelter the incompetent. I've also supported and applauded their leadership where justified.

My assertions are based on evidence from research and analysis and more than ten years of travel and living throughout the Middle East, where these events occurred and where many of the locals became good friends. I also draw upon fifteen years of military service to offer analysis that non-military authors may overlook. Also, a bit of 'gut' feel and reluctance to be politically correct.

One hundred years ago, soldiers sometimes used terms that are no longer polite or acceptable to describe natives of the involved countries. To give context of the time, I have cited the terms the soldiers used not to be disrespectful, but rather, in truth of their stories. Cultural differences then and now are explained in terms of the time. It is not meant to be offensive, but it is simply to record the feelings of the soldiers from their own experiences and observations.

I've used footnotes rather than the more common endnotes that I personally find annoying, in having to flick backwards and forwards in the text.

To help you understand the context in which this story unfolds, I've referenced some political events. My aim has been to tell a story of Anzacs, not politics. Nevertheless, outlining the complex politics of this theatre is necessary to comprehend how or why seemingly bizarre military decisions eventuated. The influences of the Sykes-Picot Agreement, the Balfour Declaration, the Declaration to the Seven, the Anglo-French Declaration and the Sultan's jihad call, as examples, all impacted the political strategy of this Sinai Palestine military campaign but they are frightfully complex; so much so that they are still disputed today. Slight mention is made here only, or this book would have been in unwanted volumes.

This is a book about the soldiers, told, at times with a bit of humour and at other times quite emotionally. The text includes real quotes from the soldiers themselves rather than invented dialogue to maintain the accuracy of their stories.

To the best of my research and knowledge, the contents of this book are true. Should you find an error please bring it to my attention. Please enjoy and share our previously under-told story of Anzac history.

Acknowledgements

The directors of the Great Arab Revolt Project (GARP); Neil Faulkner, Nick Saunders and David Thorpe are responsible for twigging my interest in what was to me an unknown chapter of Anzac and Australian history. David interrupted an entire viewing of the Australia vs. South Africa rugby test match in April 2008 to see what I could discover about Anzac connections with Lawrence of Arabia. Nil I told him; didn't I get a surprise! Then, the three of them indulged my perhaps boisterous telling during four field projects of how Anzacs had 'saved their British bacon'. To the many, perhaps 30 or 40, participants of GARP over those years who also indulged my colonialisms amidst their British and American ways, go my sincerest thanks.

John Scott, an American GARPer, then did his own conflict archaeology Ph.D thesis on the Lawrence of Arabia trail that captured the port of Aqaba from the Ottomans. He dragged me on three field projects in southern Jordan where I became photographer and historical researcher/sounding board. John and his dear wife, Kelly Pool, are great mates and helped me with early notes for this text. Thanks, guys.

And thanks to Berin and Danelle, my great kids for enduring and waiting for completion of what I have been ranting over for over eight years.

Rachel Vickery, a great friend and confidante, indulged my immersion in the early research and ongoing projects and absences. Now a part-time flatmate, she guards my treasured home while I travel the world to research and write. Thanks, Rach.

Barbara Porter and the fabulous staff at ACOR (American Centre of Oriental Research) in Amman, Jordan have hosted me on numerous occasions, the last and longest for five months at the end of 2015. Their support and help in the library and resources was instrumental in getting this book completed. Also in Jordan, my very good friends and supporters, Rana Naber, and her cousin, Reem and Raja Gargour. Rana was a senior assistant at the Australian Embassy in Jordan when we first met in 2008 and with two Ambassadors (Trevor Peacock and Glen White) gave invaluable assistance. Raja is the Director of the Royal Automobile Museum in Amman and representative of the Royal Hashemite Court of Jordan; a good friend who opened doors and introduced me to many wonderful Jordanian people.

Robyn van Dyk of the Australian War Memorial was a bright light. I am also very grateful for the staff and resources at Australia's National Library, Mitchell and New South Wales State Library, Queensland State Library and National Archives of Australia. In England, the National Archive London, the British Library and Imperial War Museum. In Israel, Mr Jacob Rosen, who I first met when he was Israeli Ambassador to Jordan, who still sends me items of interest from this campaign; the Centre for British Research in the Levant Jerusalem; and Ezra Pimental who gave me expert, and generously complimentary, guidance around Beersheba. In Syria (before the trouble), the Hejaz Railway Museum and Tourism Department staff and my guide and friend, Rawad Alaryan in Damascus; he recently escaped to Canada and safety.

Ali Baldry is another GARPer. She loaned me her London flat where the weather was so winterly dreadful in February and March of 2016 and only marginally better in the spring of May that I was able to stay indoors without distraction, and complete the draft of this text while she travelled England and Wales on her own field forensic archaeology projects.

A book like this would never have been possible if the young men of Australia and New Zealand had not taken up the call and gone to war. To them, our countries owe gratitude and our most sincere thanks; for those who came back and those who did not. They gave us a history and legend that they, and we, can be truly proud of. Thank you. And too, the vanquished; the Turkish and German soldiers, unfortunate tools of failed diplomats.

Abbreviations and Terms

AAVC	Australian Army Veterinary Corps
AFC	Australian Flying Corps
AGH	Australian General Hospital
AIF	Australian Imperial Force
ALH	Australian Light Horse
AMD	Anzac Mounted Division
ANZAC	Australian and New Zealand Army Corps
Archie	Anti aircraft fire
ARU	Army remount unit
Baksheesh	Payment for a deed; bribe
Bde	Brigade
Bedu	Bedouin tribesman/woman
BF	Bristol fighter aircraft
Bing Boys	Hong Kong and Singapore (artillery) mountain battery
Brig	Brigadier General
Capt	Captain
CCS	Casualty clearing station
C-in-C	Commander in Chief
CIGS	Chief of the Imperial General Staff
CO	Commanding officer
CTC	Camel Transport Corps
CUP	Committee of Union and Progress
Div	Division
DMC	Desert Mounted Corps
EEF	Egyptian Expeditionary Force, British commanded
ELC	Egyptian Labour Corps
FA	Field ambulance
GHQ	General Headquarters
GOC	General Officer Commanding
Gyppo	Old term for an Egyptian native
Hajj	The fifth (of five) Pillars of Islam; holy pilgrimage to Mecca
Hod	Small clump of palm trees, sometimes with water
HQ	Headquarters
ICC	Imperial Camel Corps
Jackeroo	Trainee stockman on sheep or cattle station
Jacko	Turkish soldier
Jihad	Muslim call to uprise against Infidels
Johnny Turk	Turkish soldier

Khamsin	Strong wind originating in Egypt, blew thick dust storms for days
LHFA	Light Horse field ambulance
Lt	Lieutenant
Lt Col	Lieutenant Colonel
Maj	Major
MVS	Mobile veterinary section
NAA	Northern Arab Army
NCO	Non-commissioned officer
NZMR	New Zealand Mounted Rifles
OC	Officer commanding
POW	Prisoner of war
RAF	Royal Air Force
RAP	Regimental aid post
Regt	Regiment
RFC	Royal Flying Corps, became the Royal Air Force
RN	Royal Navy
RNAS	Royal Naval Air Service
RSP	Red Sea Patrol (RN)
SAA	Southern Arab Army
Sqn	Squadron
Stand to	Bring troops to a defensive position and state of high alert
Taube	German aeroplane
Territorial	Part time soldiery; Army Reservists
Tommie	British soldier
Vanguard	Leading elements of an advancing force
Yeomanry	British mounted infantry with swords (cavalry had sabres)

Introduction

The Great War conflicts in France and Gallipoli gave Australia Anzac Day and its fighting legend. The Sinai Palestine Campaign confirmed that fighting legend and in addition, gave the world the Middle East chaos of today. The link between these campaigns was the Suez Canal.

Opened in 1869, the Suez Canal gave international shipping direct access from the Indian Ocean to the Mediterranean Sea and cut weeks off the previous journey around the African continent before this Great War. Whoever controlled the canal in wartime would have a tremendous advantage and it should have been Britain.

The British had occupied Egypt and controlled the Suez Canal since 1882. During the war, the canal was strategically priceless to Britain; essential for the supply of men, materiel and millions of Australian gold sovereigns to Europe and the Middle East. German shipping, denied the canal, was deflected down the west coast then back up the east coast of Africa before accessing the Red Sea, the Indian Ocean and their East African and Pacific colonies.

Control of the Suez Canal dominated British, German and Ottoman strategy throughout the war.

When England went to war so did Australia, New Zealand, Canada, India, the West Indies and South Africa as British dominion countries. The Anzac dominions gave the British Empire its first victory of the Great War two years after it started, at the Battle of Romani in August 1916. Twenty-five miles from the Suez Canal, the Anzac's Sinai Palestine Campaign began. Anzacs then rode, flew or drove on for another two and a half years to forge an eternal national heritage.

The Anzacs' Sinai Palestine Campaign has not previously been explained in any detail, despite its greater influence on mankind than anything from this war or WWII.

In those days, dominions were simply referred to as 'British'. Little recognition was given to national identity, even in their home countries. Domestic governments and media lived in a 'King and Empire' dreamland of subservience.

Throughout the campaign, some politicians and diplomats in England understood the need to hold the Suez Canal and provide the military resources to do so. Others, however, gave priority to the Western Front of Belgium and France while to them, Sinai Palestine became a sideshow. Political confusion often drowned military considerations and trained English soldiers were taken from the Middle East back to France, to be replaced by untrained citizens who had worn their uniforms for minutes, rather than months. This wretched replacing of experienced soldiers with shopkeepers and farmers plagued commanders throughout the campaign, extending its duration.

Only the Anzacs, led by Australia's Lieutenant General Sir Harry Chauvel and New Zealand's Major General Edward Chaytor, provided continuity from beginning to end. The Anzacs gave stability to the British led Egyptian Expeditionary Force (EEF). They provided the warrior advantage to British commanders and likely saved the British Empire.

Coincidentally, an influential group of Arabs rose in revolt against the Turks at around the same time Chauvel's Anzacs won victory at Romani. From then on, the British and Arabs maintained a wobbly alliance. The campaign was then fought on two fronts with the Jordan River giving a fuzzy demarcation line that soldiers and Arabs crossed as needed. To the west of the river was the EEF; to the east was generally Arab territory. The coordinated aim became the destruction and expulsion from Palestine of the Turks. The British eventually provided the Arabs with weapons, instructors and liaison officers including Lawrence of Arabia, money, aircraft, ammunition, food and ships to motivate the Arabs in their role. The Anzacs operated on both sides of the river, supporting the British and the Arabs.

Lawrence of Arabia is one of the most recognised names from the Great War and books about him fill libraries

Introduction

around the world. Yet he was in a campaign area that is little acknowledged. Lawrence's exploits would have been doomed had he not received Anzac support for his Arab forces but this has never been explained or acknowledged until now.

During my participation on four Bristol University led conflict archaeology projects along the Hejaz Railway in southern Jordan, called the Great Arab Revolt Project (GARP), I discovered the significance of Anzac involvement there. I also discovered an absence of quality research and writings on the Sinai Palestine Campaign in general and Anzac involvement in particular. Sure, there were unit histories, official histories and books written by soldiers who were there. Written immediately after the war, these accounts were all subject to the restrictions of the writers who only saw their patch of turf and screened by the Official Secrets Act or belief in all things holy towards the British Empire. There was little analysis of the rights and wrongs of events or persons involved until Colonel E.G. Keogh's treatise *From Suez to Aleppo*, written for military officers or students of military history in 1954, nearly 40 years later.

Desert Anzacs is not a rambling description of one battle after another intended for academics and historians. It's the story of people and their roles, thoughts, emotions and recollections of events and mates, based on their letters, diaries, postcards, books and photos. It's a story for the people of Australia and New Zealand, England and its dominions, America and the world. These are soldier stories that have been hidden for 100 years, now revealed so they can be acknowledged as part of the Anzac legend. They are in every way as meaningful and lively as those who fought in France and at Gallipoli – and many of them did fight at Gallipoli before this campaign.

Importantly, it also tells a story of the vanquished, normally overlooked when the victor tells the story. Here are Turkish and German points of view too for they were sons, brothers, fathers and friends with lives and concerns of their own. And those soldiers didn't start the war, their politicians and diplomats did.

They were just the tools. However, access to Turkish archives is prohibited so official stories don't exist. Their soldiers' personal writings are revealing.

Arabs of the time wrote no English and little in Arabic. Their stories were handed down through generations around campfire dinners, accuracy now largely lost in the passing of time.

Universally acknowledged are the soldiers who formed the Anzac tradition in France and at Gallipoli. At long last, also acknowledged are those soldiers who performed to the pinnacle of the Anzac spirit in Sinai and Palestine. Now due recognition is given to the families of those soldiers because on many, many occasions in talking with people about this book they'd say 'Gee, my great so-and-so was a light horsemen, is that where he was?'

Now they know.

Desert Anzacs tells the story of more than the light horsemen. Now we see cameleers as part of the mounted force. Australia was the only dominion country to provide airmen. Usually ignored are the logisticians and background crews, without whom the fighting men could not fight. So here we also have the story of the medics, veterinarians, armoured cars, engineers, rough riders and many others.

Gallipoli and France were septic affairs, submerged in the mud of water-engulfed trenches. Sinai Palestine was a fast, open style war that better suited the Anzac outdoorsmen. This was a new age war. Wellington and Napoleon, with muskets good for 50 yards and cannons good for 300 yards, were gone. Now, there were aeroplanes, tanks, high velocity guns, poison gas, hand grenades and armoured cars with machineguns. Many youngsters would get a horrible shock.

Because it was open and fast, there were fewer casualties than the absurd over-the-top charges from trenches; a good reason the media weren't interested and a home public denied this story. Nevertheless, all would discover that war is hell, no matter where it is fought, and one bullet finding its mark is as

Introduction

deadly as any other.

Anzacs went to Gallipoli and exposure to incompetent British generals and a determined Johnny Turk. Evacuated to Egypt, they would once more face incompetent British generals and a determined Johnny Turk as they crossed the heated sands of Sinai. They endured the worst environment Planet Earth could throw at human beings. Summer or winter mattered not; extremes of heat and cold, wind and dust, crawling and flying creatures, waterless, sickness and disease, little rest and recovery.

Harry Chauvel is one of the most brilliant leaders and military officers Australia ever produced. Now revealed is his story as commander of Anzac, British and dominion soldiers, the first non-British officer ever to do so. His story is a role model for military officers and personnel, business leaders and employees, sports coaches and players, parents and children and those who seek personal achievement and success.

Harry Chauvel guided the Anzacs through the Middle East that has been a cauldron of calamities since the last shots of the Great War. It has driven global tensions and decisions and perpetuated hatred, much more so than France and the Western Front arenas. Yet its significance has seldom been explained; it is now.

Further, this war would change world structures resulting in 'the wrecking of Europe, the destruction of four great empires (Germany, Austria-Hungary, Russia and the Ottoman) and the fatal weakening of two others (Britain and France). And it would shape and influence events in the Middle East to our own time'.[1]

Many in the Middle East and Arab world today accuse Britain and France of deceit and treachery about broken promises offered to attain Arab support and to reject the Ottoman Sultan's call for jihad. They believe they were promised the reward of an independent Arab nation. They

---

[1] Grey, J., *The War with the Ottoman Empire*, Oxford University Press, Melbourne, 2015, p. 5.

blame the accused for creating a Jewish homeland in their midst and making their lifestyles somehow worse. They blame those countries that attended supposed peace conferences after the Great War. These led to the League of Nations dividing Arab lands into mandates where supposed advanced countries provided tutelage to the undeveloped aborigines of the Arabs. They conveniently ignore the endless disputes between Arab leaders then and now.

Since those mandates, each Arab country has attained its own independence. But none have bonded to form an Arab nation of any continuance. Tribalism, sectarian divisions, ancient feuds, distrust over faith, those with the oil wealth over those without – and goodness knows how many other real or imagined issues have delivered an Arab Spring and radicalisation that has the world bewildered. Thinking Arabs and Muslims reject violence and seek a peaceful existence. What can be said, however, is that events and decisions from 100 years ago made a major contribution to the Middle East of today. This book tells some of that story to give a clearer and hopefully more balanced view of events and illustrate the forces driving the seemingly insolvable conditions of our contemporary world.

At the time, Anzacs were bystanders in the political scene. Soldiers just do their military job, made more difficult by politicians and diplomats in their throes of absurdity that ensures peace is never lasting.

Anzacs would again be called upon within twenty years.

Part One

From Home and into Sinai

1: Lose Suez…Lose the War

You must find a way or make one.
– Hannibal

Most were young, many too young to even vote. Others weren't so young. Thousands of Australian soldiers stumbled wobbly-legged off ships after weeks at sea. For most, it was their first time on a ship. They'd crossed the Indian Ocean, sailed through the Red Sea and up the Suez Canal, then between the sands of Egypt on the left and Sinai on the right. Packed into antiquated trains, they'd slung rifles and packs onto stained floors covered with dirt and trash, then dropped onto hard wooden seats with broken slats and no cushions. They bounced around all the way to Cairo.

Tent cities popped up right next to the pyramids, just as they had in gold-rush boomtowns of Australia's Bendigo and Ballarat decades earlier. They camped beside the young soldiers of New Zealand, Canada, South Africa, India, West Indies, Great Britain and the Ghurkhas of Nepal. Then they prepared for France and Gallipoli. First they had to acclimatise, continue the training these volunteers briefly started at home, practise their musketry, ready their horses and bond with new mates. Day after day, as weeks became months, they trained. It was hot, hard work. When they got leave, those naughty Australians cut loose. The market traders, beer sellers and loose women of Cairo had never before experienced such wild behaviour.

For the next few months, the young soldiers camped and trained near the pyramids, about 25 miles from the Suez Canal. Along that canal a force of garrison troops, mainly British,

Indian and Sudanese, were set to defend it. That canal became the focus of British wartime thinking in the East for the next four years. And it became the centre of attention for many young Anzacs in the desert sands, suffering heat and thirst, tribal thieves, milk and honey, among the holy sites.

But a nasty surprise lay ahead. The Turks and Germans wanted that canal.

The Suez Canal was to Great Britain what roots are to trees. British control of the canal maintained life: troops from Australia, New Zealand and India could join the horror in Europe, Gallipoli and Arabia. Shipping from the Indian Ocean and Persia to the Mediterranean was quicker and safer than around South Africa where British ships were fodder for German U-boats and battle cruisers lurking in the Atlantic Ocean. Oil (essential since the Royal Navy had converted from coal to oil in 1908) and trade could pass east to west unhindered via the canal; ships of the Royal Navy's Red Sea Patrol (RSP) and the French Navy could move freely to blockade the Arabian Peninsula and prevent food and weapons getting to the Turks; and best of all, Germany was deprived of rapid access to its East Africa and Pacific colonies.

The Germans on the other hand, wanted to eliminate these advantages. What's more, German ships, to access the Red Sea or Arabian Peninsula and Yemen, had to sail down the west coast of Africa and up the east coast. Impossible, so there were no German surface ships in the Red Sea and no interference to British and French naval operations or the oncoming Arab Revolt. Except, submarines were an ever-present threat.

On the night of 2 February 1915 sleep came easily and deep. Dreams were gentle for the unbothered and somewhat bored soldiery of His Majesty's armed forces. They were stretched along the west bank of the Suez Canal, like every other night they'd been there – on the side away from any approaching Turks and Germans. This was Britain's passive defence.

Erupting from the empty darkness, around 3am, screeching klaxons slammed the defenders' eardrums. Dormitory lights flickered to brilliance then scalded startled eyeballs. Exploding shells rained among their tents. Terror struck, dazed men fell from beds. Rifles fired outside. Men yelled, ran. Machine guns belted out death songs. In mixed states of dress, soldiers scrambled for their rifles.

The Turks had arrived.

'Stand-to!' was screamed all about. Sergeants and corporals ran quicker than startled wildlife but hastily sorted their troops into squads that stumbled, rolled, ran any way they could to their trenches and weapons. Officers gathered their sergeants and a defence organised. Shells fell and burst. But now, return fire came from the British lines to support those few sentries who had faced the start of the ruckus. Flares lit the sky. Hordes of small boats were scattered over the canal. The British shooting increased with intensity and accuracy and found their marks. Naval ships turned on their searchlights and fired their big guns and machine guns. Turkish boats hit the west bank, their occupants ran up the beach towards the British trenches – many fell; most fell. Other boats on the canal faltered, some sparked like New Year's Eve fireworks. Others disappeared below the surface. Many turned back to the east. Smoke thickened, spewed out from incessant firing. Fallen from boats and weighed by heavy clothing and boots, those who could swam and scrambled up the beach, and when they were able to stand raised empty hands above their heads, forlorn. Others drowned. Cheers rose from the defenders as the Turks beat a retreat.

Calm returned with the daylight, but the Turks had been and gone – 16,000 of them. Where the hell had they come from?

Across the bloody Sinai.

A few days later while gazing over that expanse of the Suez Canal – not that wide really, a bit over 100 yards – General Maxwell, Britain's Egyptian Force Commander, reflected on the words of Lord Kitchener, Britain's wartime Secretary for War, his cruel words: 'Are you defending the canal or is the bloody canal defending you!' It hadn't been a question.

Maxwell, a man of some administrative capacity but little

recognisable military skill, consoled himself that the Turks hadn't captured the canal, hadn't blocked it. The 'blighters' had been turned back, all 16,000 of them with their 10,000 camels, 300 oxen, heavy-wheeled artillery, pontoons and boats. Hadn't got a foothold in Egypt. Like his soldiery, he hadn't known they were there either – Kitchener's point it seemed. Maxwell was blind to the critical defence of the canal, that vital carotid artery of the British Empire. Had it been captured or blocked with sunken shipping, the sort of thing artillery can do to ships in a cramped waterway, the British war effort would have been doomed. Another Kitchener point.

The Turks, although somewhat expected, had crossed 120 miles of inhospitable, waterless desert, where tracks of an army on the move stick out like... well, like Kitchener's moustache; it stuck out so much they'd used it on recruiting posters. Unknown to be on the canal's edge at the time, they'd snuck onto the waterway in the early hours sheltered from view by a huge sandstorm that obliterated vision on both sides. Once the sand settled, the attack was on.

The sentries had done a marvellous job spotting them in the dark and raising the alarm. After 1,500 or so had been shot or captured, the rest turned and fled back into the desert. And they kept running, unmolested – passive defence being passive.

Maxwell mused. Egypt had a couple of hundred thousand troops getting ready for Gallipoli or ready to go to France. All those bloody Australians terrorising Cairo, what were they supposed to be doing? How could 16,000 Turks beat 200,000 or 300,000 British soldiers anyway?

In the lead up to this attack, the Turkish leaders, Enver Pasha and Djemal Pasha, their close advisers and the German Ambassador had set their eyes on the Suez Canal. They believed a quick descent on Egypt to capture and hold the canal, then control the sweet water of the Nile and deprive British troops, were necessary to destroy the British advantage. Threatening Egypt would also tie up a large British force outside France, a key German objective. They also hoped such an audacious move would inspire Egyptian and Indian Muslims to join the jihad[2] and rise against the British. The concept sounded persuasive to non-military minds.

General Liman von Sanders, Head of the German Military Mission to the Ottomans, was a lonely, dissenting voice. He assessed that the British military and naval presence in Egypt was too strong to allow the Turks to capture and effectively hold any ground. He realised the immense logistical requirements to send a force of sufficient size across the desert and capture the canal was absurd; instead, he wanted to block it by sinking ships in its narrow and shallow waterways. Under pressure from the Turks and his Ambassador, he was told to 'pull his head in' by the German High Command. In anguish he said, 'I have stated that the undertaking was condemned to failure from the beginning. Egypt cannot be taken by 16,000 Turkish troops'.[3]

The attack had failed. But German and Turkish focus continued. A bit over a year and a half later they would have another go, but into the face of Anzacs and their Australian commander. In the meantime, British and French political and military strategists came up with the dreadful Dardanelles naval campaign in March 1915 to assist their Russian allies. That failure was quickly followed by the Gallipoli disaster from April to the end of 1915. Nor was there anything positive or joyful for the British on the Western Front where stalemate prolonged.

So ended 1915; the Turks jubilant and the British despondent, seeking a victory, any victory. A British shake up seemed overdue. New generals were needed.

Australia had one.

---

[2] The Ottoman Sultan, in November 1914, had called a *jihad* or holy war against the infidel Christians; the British, French and Russians. They hoped that Egyptian and Indian Muslims would also join the *jihad* against the British.

[3] Von Sanders, L., *Five Years in Turkey*, p. 43.

2: Desert Commander, Harry Chauvel

Nothing great was ever achieved without risk.
– Machiavelli

He was a bright Australian kid. The sort of bright that made life work for him; the trees, the rivers, the paddocks, the stars, the sun, even the clouds and rain; where he grew up and built his communion with the outdoors. Maybe he was bright at school too, but his great loves were the bush and horses, the army and his family. Harry became one of Australia's greatest military leaders never to have his story brought to life until now.

Before the better known of our generals like Blamey and Monash, Harry was the first Australian military leader promoted to lieutenant general. He was knighted twice in the field for outstanding leadership; he was the first non-British officer to command British troops (an unheard of honour that horrified traditional caste-conscious relics of the empire) and appointed by a British Commander-in-Chief; the commander of all Anzac troops in the Middle East campaign. All who knew him respected this quiet humble man with a no-nonsense manner, based on courtesy and politeness.

He was born on 16 April 1865 on a rambling cattle run at Tabulam on the banks of the Clarence River in northern New South Wales. The newborn Henry George Chauvel was immediately and thereafter known as Harry, with a backyard of rolling hills, timbered lands full of Australia's unique wildlife, creeks and the river where the cattle and horses ran wild.

His ancestors had served in British and Indian armies until Harry's grandfather, Charles Henry Edward Chauvel, fresh from the British Army in India, was among the tide of arrivals in the colony of Sydney in March 1839. In 1848, Charles bought Tabulam, a station that carried 7,000 sheep and to which he soon added the best stud cattle from England and Scotland. Adjoining properties were bought and by 1857 Tabulam consisted of 51,000 acres with something like 8,000 sheep and 2,000 cattle. Harry's dad took over Tabulam when grandad retired to Sydney around 1858.

At the age of eight, he was packed off to boarding school near Goulburn then off to Sydney Grammar School a year or so later. There, he rubbed shoulders for the first time with Banjo Paterson, who became a famous Australian journalist and poet. He won prizes for shooting and jockeyed in many a horse race, winning more than a few. In 1881, Harry, now a medium height, wiry 17 year-old, returned to the homestead and a pastoralist life. But he wanted a military life. That would mean having to go to the Royal Military College in Sandhurst, England, as there wasn't any in the colonies at this time. Together with tough rural living, droughts and poor family finances, Harry missed out on Military College altogether. Skirmishes in the Sudan and threats of Russian activity in India saw father Charles granted permission to raise two cavalry troops around the Clarence River. Dad became a captain and sons Harry and Arthur became second lieutenants. The Upper Clarence Light Horse, as they became known, weren't sent overseas. But times got tougher, dad had to sell Tabulam and they moved to Canning Downs in Queensland to a smaller property. Harry didn't hang around for long and skipped off for a year or so in Europe, travelling, meeting and greeting, visiting art galleries and music theatres. But he also maintained his military interests, visiting British regiments and events, including the military academy at Aldershot in the presence of Germany's Kaiser Wilhelm II. He returned to Canning Downs in 1890 and his military life. Giving up his New South Wales military commission, Harry moved to the Darling Downs in Queensland. Together with his mates he established a Queensland Mounted Infantry troop and was commissioned as a second lieutenant in the 1st

Company Darling Downs Mounted Infantry in January 1890 at the age of 25.

In April 1891, the shearers were striking against the pastoralists amid threats of scab labour. An inadequate-sized police force was called upon to keep harmony and the Mounted Infantry were called upon to give aid to the civil power. Harry made a name for his coolness under pressure and for the discipline of his troopers. His small band of police and troopers had to escort four shearer troublemaker prisoners and non-union labourers through a larger group of wild, armed shearers, threatening violence. With the troopers given the order to load, Harry escorted his party through the ranks of striking shearers with rifles pointed outwards to both sides and calmly made their way through the strikers and completed their task. The inexperienced 26 year-old military officer learnt the value of discipline and training in the period of the Shearers Strike.

For five months, his young troopers watched over the Shearers Strike. When they were not required for policing work, to relieve the boredom, they'd chase emus and pluck their feathers on the run. Harry related that troopers of the Gympie Company of the Darling Downs Mounted Infantry started putting plucked plumes in their hatbands in memory of the glory. The craze caught on with the mounted troops in other areas and the distinction was extended to companies in South Australia and Tasmania in 1903 and to the whole of the light horse in 1915 – the feathered emu plumes became the symbol of the Australian Light Horse as they still are today.

Strike over, Harry returned to rural life and made another name for himself as a breeder of fine horses, a rider at race meetings, managing the Canning Downs property and as a respected military officer. On 9 September 1896, Harry swapped grazing and rural life for full-time soldiering; his true passion. He was appointed a regular army officer as adjutant, Moreton Regiment, and so began his outstanding military career.

Eighteen years later, Harry, and many of Australia's youth, would be called by the drums of war.

3: Getting to War

Courage is not the absence of fear. It is the ability to master fear.
– Wayne Bennett

A decade and a half before that attack on the Suez Canal, the twentieth century had exploded with spectacular advances. As a result the destructive impact of war was greatly increased with new technologies. Soldiers had always suffered, but this time massive civilian populations would too. Aircraft could be mounted with machine guns and drop bombs, submarines could fire torpedoes, and poison gas could burn skin and attack the mind. There was heavy artillery, aircraft carriers, tanks, armoured cars, rapid-fire machine guns. Military medicine was revolutionised. For the first time in warfare, medics had the ability to treat wounds from new high velocity weapons and to treat infections. Field hygiene reduced sickness and disease. Instead of putting wounded men onto rough riding camels or horses, trucks could be used as ambulances. Trains could evacuate the wounded and reinforce manpower in large numbers. Logistic payloads were bigger and more efficient on trains and trucks than on camels and horses. Aeroplanes could drop supplies and propaganda from above. Wireless communications would speed the passage of information.

Meanwhile, Australia's six colonies had become a Commonwealth with Mr Edmund Barton as the first Prime Minister. The Right Honourable John Hope, 7th Earl of Hopetoun, became the first governor-general.

At that time the Middle East was ruled by the 600 year-old Ottoman Empire – once a massive empire coming after, yet rivalling in size and influence, the mighty empires of Rome and

Greece. Starting in Anatolia in Turkey, the empire flowed into Egypt, Libya, Tunisia, Ethiopia, the Sudan, the sands of Sinai, the plains of Palestine, Syria, Turkey, down the Mesopotamian rivers into Iraq and Persia (Iran today), the Arabian Peninsula, Yemen, Greece, the Balkans and through eastern Europe until it banged on the gates of Vienna. It controlled the Black Sea, the Aegean Sea, the Red Sea and much of the eastern Mediterranean Sea. In triumphant times, the Ottomans controlled the massive trade routes through the unique land junction that joins Asia Minor to Europe at Constantinople (Istanbul today). It was an empire controlled by wealth and military strength. It was a palette of races, religions, cultures, customs, skin colours, languages, gastronomy, architecture and clothing.

The Ottomans had ruled over the construction of the Suez Canal, then the greatest waterway of the planet. But the Ottomans couldn't meet their loan repayments as a result of bad economic management and the British took control of the canal with a military presence in 1882, finally declaring it a Protectorate at the beginning of this Great War.

Towards the turn of the nineteenth century, the Middle East was reasonably stable under the authoritarianism of the oppressive Ottoman Sultan Abdul Hamid II, who came to power in 1876. He had the title Caliph and led the Caliphate as the head of the Islamic religion. The Ottomans ruled their world from Constantinople while overseeing the holiest Muslim sites of Mecca and Medina, Jerusalem and Damascus.

Sharif Hussein bin Ali, a direct descendant of Mohammed, was leader of the Hashemite tribe in the Hejaz region (the western region of today's Saudi Arabia). He was protector of the religious centres of Mecca and Medina but had been 'extradited' to Constantinople in 1893, along with his three sons and wife, by the Caliph Sultan. The Sultan believed Hussein was influencing the Arabs of the Hejaz to disobey the empire. For fifteen years, Hussein educated his sons in religion, political, military, social, economic, and affairs of state, in preparation for an eventual return to the home sands and squabbling tribes of the Hejaz.

However, trouble brewed throughout the empire. Rebellious Arab discontent had grown so much that, under pressure from the revolutionary Young Turks of the Committee of Union and Progress (CUP), the Sultan returned Hussein to his homeland in December 1908. Supposedly to quell that discontent, they appointed him Emir of Mecca, defender of the Islamic holy sites in Mecca and Medina with responsibility for the safety and comfort of Hajj pilgrims.

The Sultan was a despot and brutal. Others sought his political overthrow. The CUP led by the Young Turks did just that. In 1909, the Sultan was deposed and replaced by his docile brother, Mehmed V, and the Young Turks took over. Their aim was to end multiracial favours and 'Turkify' the empire.

No empire lasts forever. Under the inept rule of the Young Turks, the Ottomans lost wars and their Libyan territories to Italy (1911), Tunisia to France (1912) and the Balkans (1912–1913), and the British had taken over Egypt in 1882. The tribes of the Arabian Peninsula were disgruntled and wanted independence. The Russians had re-occupied their Black Sea coast. Political turmoil bubbled in Damascus, Beirut, Cairo and throughout the Arab world after the Young Turks executed by hanging 25 Arab notables in their endeavours to Turkify the government and remove non-Turks. The coffers were empty. Its military was disorganised and ill-equipped. The empire was broke and continued to crumble. The Ottoman Empire became known as 'the sick man of Europe'.

Optimistic Young Turks courted Britain, France and Germany for support, yet alienated their Arab citizens: strange thinking and behaviour when stability was needed. All that remained of the once mighty empire was Turkey, Syria, Palestine, the Arabian Peninsula and tribes of the unwanted Sinai who cared nothing and responded little to the Turks. The Ottoman army had been decimated; officer strength had been decimated; soldier quality had been decimated; the air force had nothing and there were no ships in the navy until two

German cruisers and their crews became 'Turkish' in a swift deal. Had it not been for German assistance with training and re-armaments during 1913 to early 1914, things would have been as desolate as a Papal wedding breakfast.

The Arabs of the Hejaz considered revolting against the Ottomans. In 1912 and again in early 1914, Emir Abdullah, second son of Hussein, approached Lord Kitchener to sound out British support for such an uprising. No war, no support, was Kitchener's reply. But encouragement was given should war erupt.

All that was needed was a war, and the Germans came to the party.

The Great War finally got going after Germany trampled over Belgium and France. Great Britain retaliated by declaring war on Germany on 4 August 1914. Australia's Government received the news like free beer at a wedding. The Governor-General Sir Ronald Munro-Ferguson blurted:

> There is indescribable enthusiasm and unanimity throughout Australia in support of all that tends to provide for the security of the [British] empire in war.[4]

As a dominion, Australia was obliged to go to war with Great Britain, but our government was able to determine to what extent it would contribute. A few days earlier, in anticipation of war, the government of Prime Minister Joseph Cook pledged to Great Britain: '20,000 soldiers and all our naval ships'.[5] On the declaration of war the Minister of Defence, Senator Millen, screeched: 'Australia is no fair-weather partner in the Empire'.[6] And the Leader of the Opposition, Mr Andrew Fisher, who became Prime Minister at the election on 5 September, shouted at a public meeting: 'The last man and the last shilling will defend the Empire'.[7]

'Bloody beaudy!' was the cry of innocent, maybe naive,

---

[4] Scott, E., *Official History of Australia in the War of 1914–1918, Vol XI*, p. 13.

[5] Scott, p. 11.

[6] Scott, p. 16.

[7] Scott, p.16.

Getting to War

youthful exuberance. Men enlisted in droves, faster than sheep on the run could jump fences. Age was no barrier to enthusiasm. Many youngsters without parental consent forged their ages to enlist and fathers with sons joined the lengthening queues, lest there be no uniforms left.

The Australian population was jubilant, euphoric even. There was no dissent from any side of politics, not from civil rightists, not from pacifists. Everyone supported their heritage. It was our duty to the empire. Never having seen the reality of war, Aussie boys volunteered with quick-fire tenacity. Cook's offer of 20,000 became 400,000. Bands played and children sang.

Australia's ladies weren't to be left out:

> The proposal to form a corps of Australian women volunteers to assist in all patriotic movements, and to be drilled in the use of the rifle for home protection is looked at cheerfully by representative military authorities who regard the idea as laudable.[8]

One officer was so bold as to suggest:

> Personally, I think it is a very good thing that women should hold themselves in readiness for their self-defence. Every woman in the country should be taught to shoot. They should also be taught to swim. If women could handle firearms there would be less burglaries.[9]

The women didn't need to swim their way abroad. In fact, the only women in uniform to go were nurses of the Army Medical Corps. More than 2,000 Australian nurses served overseas during the war. One year after war's beginning there were around 620 nurses serving in Europe, Egypt and New Guinea.[10] Many wives of officers in Cairo provided recreational facilities as an alternative to the bars and brothels for men on leave and assisted the overworked nursing staff to care for wounded soldiers in the hospital system.

---

[8] Stewart, D., Fitzgerald, J. and Pickard, A., *The Great War: Sources and Evidence*, p. 171, quotes the *Sydney Morning Herald*, 12 September 1914.

[9] Stewart, Fitzgerald, and Pickard, p. 171.

[10] Stewart, Fitzgerald and Pickard, p. 172, quotes *Australian Nurses Journal*, 16 August 1915.

In August 1914, three brothers, keen to do their bit for the war effort, sat around their kitchen table listening to the wireless report of Germany's rampage across Europe: Robert, Albert and the fifteen year-old Eric Bolton-Wood. Young Eric wrote himself a letter of permission, signed it as his mother and trotted off to enlist. 'How old did you say you are, young fella?' demanded the recruiting sergeant. 'Nineteen, just like it says in the letter, sir,'[11] came the eager reply. Off to the doctor went Eric, then into a uniform. He ended up in Egypt with the Imperial Camel Corps (ICC) then served throughout the Sinai Palestine campaign. This cheeky pup, despite being much younger than most of his mates, was promoted through the ranks of corporal (aged sixteen), sergeant (aged seventeen), staff sergeant (aged eighteen) and warrant officer class 2 by his real age of nineteen – that's an extraordinary accomplishment for any man, even in war.

As the war's casualties mounted, more men were needed. On 17 March 1917, in the tiny country town of Walpeup, northwest Victoria, fifteen-year-old Harold Bell, weighing in at eight stone with rocks in his pockets, wrote a note to his mum and dad telling them he was going to Queensland to do some jackarooing. Harry could ride bareback, work endlessly with limited food and water, shoot the eye out of a duck on a creek and find his way through unknown countryside – he was a natural for the light horse, his dream. Bag packed and note on the sideboard, Harry snuck out to the recruiting office.

Harry filled out his paperwork saying his name was Harry Wickham, aged 21, with the next of kin his uncle Thomas Bell of Walpeup, Victoria. The doubting recruiting officer asked, 'You look a bit young. Got a birth certificate?' Harry creatively replied, 'No sir, it was lost in a fire few years back', to which the recruiter responded, 'What about your parents?' 'Died in the fire, sir. Got my uncle but we don't talk much'. Passing this inquisition, Harry Bell became Trooper Harold Wickham of the

---

[11] www.anzacday.org.au/justsoldiers/WO2 Noel Bolton-Woods.

4th Light Horse Regiment.¹² He arrived in Palestine in August 1917. In September he was allocated to the machine gun section and given his horse, a Hotchkiss machine gun and a second horse as packhorse – two horses, double the excitement, thought Harry. And so another youngster joined Australia's legend.

Mateship, that driving force of Australia's heritage, saw the menfolk of country towns and villages enlist as groups. City kids fell in with suburban 'cousins', forging camaraderie. Newly formed military units had a home-grown tag. They were boys from the bush or outdoor workers in trade or labouring; Man from Snowy River types. All were thrilled to don the uniform, grab a rifle and march the town streets to bugle and drum; and the crowds cheered.

They came from the only inhabited continent on earth never to have had its beaches, rivers and lands run red with the blood of war. Their battle experience was the odd pub brawl or tiff over a young maiden; then a handshake and another beer, conflict forgotten. But many had learnt to ride and shoot before they'd opened a schoolbook; they could live outdoors for weeks at a time, observe, hunt, work alone, work in a team, live off the land, adapt to new surroundings. They could rough it in adverse weather; heat, cold, rain and snow, and use their initiative to make the decisions needed to get a job done. In this harsh frontier, a man's toughness and mateship meant survive or perish. But all those skills and talents that seemed so natural, weren't as conducive to the battlefields of France and Gallipoli as they would be to the Sinai and Palestine. The Kiwis were of much the same stock.

Few were real soldiers. Australia hadn't yet developed a sizeable, full-time army in the decade or so since Federation, nor had New Zealand. These lads were volunteers for King and country with dreams of wild adventure priming their enthusiasm. Trooper Robert Bygott was a migrant from England who some years earlier had trotted off with his brother to the

---

[12] www.lighthorse.org.au/personal-histories-harold-thomas-whickam.

sheep station Inniskillen near Blackall in Queensland. Both raced to the recruiting office in Brisbane. In his memoir *With the Diggers in the East*, Bygott recalls, 'Of course, the fellows who had come into camp the previous day were old soldiers now, 24 hours old'.[13]

Australians and New Zealanders were egalitarian; full of adventurous spirit born of the need for mutual reliance to survive in harsh lands, without military tradition in their new nations. They were not like the professional British Army with centuries of tradition, pomp and pageantry, hierarchy and inflexibility of class and rank structure. Ability in battle would be the yardstick for promotion among officers and would allow good soldiers to become officers – near impossible in the caste-conscious British system.

The Australian Government sent its troops off with the following cablegram:

> Australia places her troops unreservedly at the disposal of the [British] War Office, for employment in any theatre of war; to be under the tactical and disciplinary command of any general appointed by the War Office but Australia reserves the right to administer, pay, equip, clothe and feed her own troops.[14]

Little did our government realise that most of those British generals had never seen battle and would be replaced too late for the many dead and injured whose lives they'd controlled.

During the war, our army was a total volunteer force of over 400,000 men from our overall population of five million. Even with the growing number of casualties as the war dragged on over the years, two referenda for conscription were defeated and volunteers made up the numbers. Britain on the other hand, had a total military force of 4.9 million of which 2.4 million were volunteers and 2.5 million were conscripts.

Harry Chauvel had spent time on a detachment to England

---

[13] Bygott, Tpr R., With the Diggers in the East, private diary provided by Mr Bob Bygott (son) from private family records.

[14] Hill, A.J., *Chauvel of the Light Horse*, p. 45.

in 1897 where he served for a while with infantry and mounted infantry, his only training at officer level. There were no officer colleges in Australia and he was now too senior to attend colleges in England and India. In 1899, most Australian attention was on Federation; but a storm was brewing in South Africa that became the Boer War. Harry was given command of one of the two companies of the Queensland Mounted Infantry and set sail from Pinkenba Wharf in Brisbane on the *Cornwall*, which arrived in Africa on 13 December. After a year's active service, he was promoted lieutenant colonel, appointed a Companion of the Order of St Michael and St George (CMG) for outstanding leadership of Australian, British and Canadian troops, with a Mentioned-in-Dispatches (MID). Harry and his troopers went home with the priceless active experience of leadership, horsemanship in rugged country, fighting in a harsh land, combating guerrilla-style tactics, essential training and discipline in soldiering.

Harry reveled in soldiering. He was an avid student of the Arab horseman warrior Saladin, the noble Muslim leader who demonstrated greater chivalry and compassion than some of his brutal crusader opponents. Harry also absorbed the tactics and leadership of General J.E.B. Stuart from the American Civil War. He became the Chief Instructor of Light Horse Schools and the Supervisor of Infantry Schools throughout Australia. In 1906, now aged 41 years, he married a local Brisbane woman, Sibyl Jopps. First son Ian was born in 1907, Edward in 1909 and daughter Elyne in 1913. Family life was another dedication of Harry's and he was a glowing model for others to follow in balancing family and career. During this time he was confirmed lieutenant colonel and served a few staff postings. In July 1914, he achieved his goal to go to London and the Dominion Section of the Imperial General Staff at the British War Office.

As war approached, Harry wanted to command infantry in France. Instead, he was kept in England to the year's end when he was sent to Egypt to command the Light Horse Division. He figured that was a good alternative. Gallipoli would intervene, where he served with great distinction, was promoted and awarded Companion of the Bath (CB). But Gallipoli, then France, was the beginning of Australian and New Zealand disdain for Rule Britannia. It was this disdain that eventually set the two young nations on a long road to independence and strategic reliance on America.

The Germans realised their threat to the Suez Canal would tie up a serious-sized British force that couldn't be deployed against them in France – and all the better if Ottoman, not German, troops provided that threat. But further, they wanted the Suez Canal, the Middle East and Asia Minor with their trade routes and to give access to their projected empire.

The British understood they had to defeat Germany in France to save the homeland and free Europe from oppression. Confusing their strategic thinking was the distraction of conflicts away from Europe in Egypt, Palestine, Mesopotamia, Salonika, Africa, New Guinea and the Pacific. These became known as sideshows but the canal had to be saved. British resources were slim. Politicians had vastly divergent views and making decisions didn't come easy.

From the moment war was declared and shots were fired, France (referred to as the Western Front) became the priority battlefield for the British and Germans. One group of British politicians and military officers believed that by winning campaigns in the sideshow theatres quickly, greater attention could be focused on France and Germany. They and those who recognised the importance of the Suez Canal and the Arabian theatre became known as 'easterners' while those whose dominant thoughts were on France were called 'westerners'. Easterners and westerners understood it was imperative to hold the Suez Canal and the Red Sea to keep the ships flowing. They simply couldn't agree how this should be done.

Early in the war, a quick victory in the East against natives of the Ottoman Empire was anticipated. This had allied nations dividing up the spoils long before victory. Russia would have Constantinople and Armenia; France would have Syria and

Lebanon; Italy would have Adalia; the Greeks their seaboard; and Britain the Ottoman lands of Palestine, Mesopotamia and Transjordan. This would leave the Turks with a small patch of land based around Anatolia, their place of origin. This became the substance of the Sykes-Picot Agreement of May 1916. In November 1917, the British, supported by the United States, through the Balfour Declaration said Palestine could be the homeland of the Jewish people. Sykes-Picot and Balfour conflicted with Sharif Hussein's dream of an Arab nation with him as king.

There were then a few religions to consider: Christianity, Judaism and Islam. The eastern campaign was to be fought in Muslim controlled lands, amid everyone's holy sites. Politicians, military officers and sheiks weren't sure how non-Muslim troops should be used in these lands while world Jews wanted Muslim-controlled Palestine to be opened and settled as their homeland.

There were tens of thousands of Muslims in the British-led Indian Army and Britain could not afford to have them side with the Muslim Ottomans. Muslim Egypt was occupied and provided food, water and a huge native workforce to support the British campaign and should not be antagonised. Strategic decisions were not easy and made harder through competing political, military and religious priorities.

Simultaneously, among parts of the British Government and military command, vociferous opposition to any campaign in Arabia grew. The focus of the war effort had to be against Germany on the Western Front and not upset Indian and Egyptian sensitivities. These westerners were influential and drove the early phase of the Great War and British strategy.

Opposing views proliferate in politics and war, often as boisterously as opposing fans at a crucial football match. Lord Kitchener[15] recognised the necessity of the Suez Canal and the

---

[15] Kitchener had served most of his adult and military/political life in the Sudan, Egypt and fought in the Boer War. He well understood Arabic people and the strategic vitality of the 'east'. Winston Churchill went some way towards a similar policy.

Desert Anzacs

Red Sea and their susceptibility to interference with shipping and supplies from India and Australasia and the Mesopotamian treasures and oil. The easterners also recognised the need for the alliance to support Russia when necessary.[16] Additionally, tying up Ottoman troops in the east would prevent them descending on the Western Front and increasing the possibility of defeat against a superior Turko-German force. Harry Chauvel, Eric Boulton-Woods, Harry Wickham, Bob Bygott and all the other Anzacs became pawns in the east/west struggle.

In reverse, the easterners' thinking appealed to the Germans who wanted to keep as many British troops away from France as possible. Yet they also wanted to win in the east with an eye to their own imperial expansion. The Germans would encourage their Young Turk protégés to attack the Suez Canal; capturing or blocking the canal became their aim.

But officers of the British War Office kidded themselves that the near waterless wasteland of Sinai was such a formidable barrier between Palestine and Egypt that no large force could cross it. Seems they didn't believe the Turks could do what the armies of Rome, Alexander the Great and Napoleon had done. A passive defence under the command of General Maxwell sat troops on the western side of the canal, hoping that the naval ships could shoot any Turks approaching through the desert. That didn't work too well.

So now the stage was set for the Entente Allies of Britain and its dominions, France and Italy, to take on the Central Powers of Turkey, Germany and Austria in the predominantly Muslim lands of the indigenous Arabs.

---

[16] This was the basis of the poorly planned Dardanelles naval and Gallipoli land campaigns that detracted from their reputations and lost them major political and public support.

4: The Good, the Bad and the Arabs

There is no innocence in war, only degrees of guilt.

On the basis of the author's birth origin, we'll call "the good" the British and the dominions. This meant, however, that when troops of the dominions succeeded, hometown Englishmen thought their farmers and shopkeepers had done a marvellous job.

Before the Great War, young Englishmen for centuries had excelled at conquering then controlling natives of far-off lands. In fact, they created an empire upon which the sun never set. They then created native labour forces or put them into military service and used them to conquer even more natives. In Egypt during this war, hundreds of thousands of locals made up the Camel Transport Corps (CTC) that became the resupply system. The Egyptian Labour Corps (ELC) became a construction force for the railway, water pipeline, telegraph and mesh roads.

The Union Jack is as well known around the world as fish 'n chips are in England. King, country and the Union Jack were thought to be supreme over their colonies but it's quite likely that had the dominions not joined the war, England and its empire would have been no more. But they did join and the empire continued.

The good formed the Egyptian Expeditionary Force (EEF) under British command.

However, Great Britain had not fought a sizeable land war against a sophisticated European enemy since the Duke of Wellington whipped Napoleon at Waterloo 100 years earlier. They had really contested only minor skirmishes such as Crimea in the 1860s and the Boers in late 1800s to early 1900s, or expeditions against the spears and arrows of natives in Australia, Africa, India, the Pacific and Asia. In this way, they had created an empire without a foothold in Europe. It's no surprise that many of Britain's numerous generals at the start were over-promoted. Few had seen the anger of shot or shell or held field command in their entire careers. From middle ranking officers to generals, many were promoted through the old-boy network or on the basis of long service. Little credence came from battle experience, ability or graduation from officer school. Officers were often assessed on their family connections, promoted to win favours and friendships, their dress, social class, known colleagues and manner of speech, rather than on military achievements.

Within the first eighteen months of the Sinai Palestine campaign, six very senior commanding generals (Maxwell, Murray, Lawrence, Wiggins, Dallas and Dobell) were removed and recycled to England. In addition, Winston Churchill and General Ian Hamilton made (temporary) career changes after their misguided efforts in the Dardanelles and Gallipoli campaigns.[17] General Townsend's poor leadership and command failures in April 1916 sent his 10,000-man army to destruction, death or capture at Kut in Mesopotamia through his arrogant authoritarianism.

British psychologist and author (himself a former Royal Engineer officer of nine years) Professor Norman Dixon MBE, when assessing the performance of British generals (including those of the Great War) wrote:

> The leaders of such armies were chosen from corps of officers who were not recruited primarily for prowess or intelligence, but because they conformed to certain social criteria. They, for instance, had to be noble, or to profess a certain religion, or, where nobility was not a passport to rank, to belong to the appropriate class or

---

[17] Lord Kitchener was also instrumental in those failed campaigns but he was drowned when the ship he was sailing on to Russia ran into a German mine and sank, coincidentally on the day the Arab Revolt that he had supported began on 5 June 1916.

caste. This is why successful generals when they emerge appear to be freaks or mavericks.[18]

A couple of the successful generals who influenced the Anzacs included Major General W.R. Birdwood (later Field Marshall Lord Birdwood), a British officer given command of the Australian and New Zealand force in December 1914. Birdwood remained in England and was commander of the Anzacs in the Eastern Front, never to visit the scene. General Edmund Allenby (later Field Marshal Viscount Sir Edmund Allenby) took over the command of the EEF from June 1917 and led it to final victory. Chetwode of the cavalry and Salmond of the Air Force were champions.

There was a small corps of capable and competent officers and senior non-commissioned officers (NCOs) who held things together. Some had served in the Boer War over a decade earlier. Some served with Australians and understood their peculiar colonial characteristics – for Australian soldiering was as different from British soldiering as the kangaroo is from a wombat – but their numbers were thin until the war extended and officers gained battle experience and capable leadership.

In their pomp and pageantry, British officers spent much time complaining that the ruffian Anzacs refused to salute them. General Birdwood responded that the citizen soldiers 'will at times trouble the authorities but they will trouble the enemy more, which is indeed their purpose'.[19] And indeed they did.

There was a nucleus of regular Tommie soldiers. The territorial and conscript soldiers from the crowded towns and cities of England, Wales, Scotland and Ireland were mainly farmers, shopkeepers, millworkers, miners and tradesmen. They were given no scope for initiative, rarely promoted to officer rank and rarely informed.

The distinction between the British officer and the soldier was a hair's breadth removed from the Indian caste system.

---

[18] Dixon, N., *On the Psychology of Military Incompetence*, p. 12.

[19] Hill, p. 65.

Private Rolls, a British armoured car driver, eventually worked with Captain T.E. Lawrence (Lawrence of Arabia) for two years and notes that Lawrence was quite an exception:

> In the [British] army, a man's clothes are so often the only sign of his authority. In our corps, as in others, orders were snapped out like curses, and salutes to officers were exacted in the fullest measure. Lawrence's orders were directions, and he cared nothing about saluting, except that he preferred to dispense with it. Instead of an order, he usually seemed to raise first of all, a question for discussion. Some, whose opinion had never been sought by an officer before, looked dumbfounded at him.[20]

Yet, this hapless and downtrodden soldiery regularly demonstrated courage and tenacity. In Palestine, they were a youthful fun-loving lot in much the same way as the young Anzac volunteers. Dominion troops who fought with them admired them, while generally lamenting their officers.

Little has been written about the Indians and little acknowledgement given them: their contribution is scarcely recorded. Their ranks were a cultural and religious blend of Sikh, Hindu, Buddhist, Muslim and Christian. Their lancers (cavalry with lances rather than swords), cameleers, infantry and artillery, with indigenous and British leaders, were well disciplined and although initially untrained and inexperienced, as their numbers swelled towards the latter stages of the campaign, they got the hang of it and fought to good effect, earning the praise of Anzac and British colleagues. Towards the later stages of the campaign, Indian artillery and infantry units replaced many of the experienced British troops that were recalled to France.

From the early days of transportation, emancipation and free settlement, the youth of Australia and New Zealand were independent operators. Men were equal, almost. Class snobbery didn't really exist, other than from the migrated upper class of

---

[20] Rolls, S.C., *Steel Chariots in the Desert*, p. 133.

English society. Formality and rank meant little in an outdoor environment where an owner got his hands just as dirty as the drover, ringer, axeman and builder. Yet, everyone knew who was boss. And the boss knew he had to look after his men. Welfare and transition to battle came easy in a classless society.

What started as a tiny, full-time army rapidly exploded into hundreds of thousands of enthusiastic and dedicated men, with tens of thousands ready to fight this Sinai Palestine campaign. Soldiering came easy to the tough outdoorsmen who always found the time to grumble about conditions, until a mate banged him under the ear and they all got on with the job.

On the basis of their nationality and opposition to the Allies and the Anzacs, we'll call the Ottomans "the bad" – commonly known as Turks – plus the Germans and Austrians. History seldom tells the story of the vanquished as usually only the victor gets to tell his tale of supremacy, somewhat like the story 'until the lion learns to speak the hunter's story will be believed' (anon). But it's worth knowing, as they too were sons, husbands, fathers and brothers who fought with passion for their cause.

The Young Turks realised they needed to resuscitate their 'sick man of Europe empire' but uncertainty prevailed. War in Europe appeared to be coming and the empire was in no state to participate; neutrality seemed the best option until strength could be regained. To speed the regrowth, military missions were sought: a naval mission from Britain under the direction of Admiral Arthur Limpus; the Frenchman General Baumann was head of Turkish police; the German General Otto Liman von Sanders became head of the Military Mission in December 1913.

As the Great War broke out, Turkey, still moribund, remained neutral until siding with the Central Powers in October. Now siding with Germany, the ramifications of politics, religion and the military would descend on the Turks. Sultan Mehmed V, at the instigation of the Young Turks, decreed his jihad to Muslims worldwide in November 1914:

> Know that our state is at war with the Governments of Russia, England and France and their allies, who are the mortal enemies of Islam. The Commander of the Faithful, the Caliph of the Muslims, summons you to the Jihad.[21]

This jihad scared the socks off British officials, fearing their vast Muslim populations in India and Egypt might follow the call and rise against them. Such fear had an overpowering influence on British policy in the East.

But the Ottomans, with scarce resources, financial ruin and a depleted soldiery, engaged operations in the Dardanelles, Gallipoli, the Russian front (the Caucuses), Salonika, Romania and Mesopotamia. And strangely, they compounded this by accelerating the fire of Arab discontent.

The ethnic and cultural groups of the Turkish army provided many colourful exchanges within the ranks, with much ill will between Turks, Arabs, Africans and eastern Europeans. The soldiers were unprepared for another war after the Ottoman conflicts from 1911 to 1913 and after they lost 80-90,000 comrades at Gallipoli. Their ranks were thin and undesirous of more war; but they were obedient.

Colonel E.G. Keogh describes the Turks this way:

> The units raised in Anatolia would have been good fighting material had time and money been spent on their training. Some 95% of the rank and file were illiterate and of low intelligence, their mastery of modern weapons was defective and their knowledge of tactics was elementary. However, they were brave and sturdy, and they possessed a remarkable capacity for living and fighting under the most adverse conditions.[22]

Von Sanders gave colourful descriptions of the Turkish Army at various times, this one early in the campaign:

---

[21] Barr, J., *Setting the Desert on Fire*, p. 3.

[22] Keogh, Col E.G. *Suez to Aleppo*, p. 17.

> A mental depression had taken hold of the entire corps of officers. Some officers had not received pay for six or eight months. The men had not seen pay for years, were under-nourished and dressed in ragged uniforms. Enver [self appointed Minister for War and army commander] possessed neither experience, judgment nor training to properly settle important questions of German collaboration. In the provinces, the strength, particularly of infantry organisations, had been very variable. In the summer of 1914 [May] I found companies with no more than 20 men.[23]

In a short time, however, Jews and Christians previously obliged to undergo military service became Turkish lackeys. Uniforms and weapons taken, they became the labour battalions; just right for building roads, backpacking stores, resupply duties, medical evacuation, providing water, loading and unloading stores from train to truck and back to train at rail gauge changes, digging trenches and fortifications, filling sandbags and a multitude of unpleasant labourings in the heat.

Throughout this campaign, Turkey was short of everything needed for a fair dinkum war: ammunition, artillery, signal and engineer equipment, even good food, uniforms, medical supplies and transport. Morale was never great but Johnny Turk fought resolutely until the very end. He was especially competent in retreat to prepared positions and defence against a rapidly advancing force.

To compound the military malaise, Turkish supply lines from Europe and Turkey into Palestine to support their soldiers were a disaster. Railway lines were incomplete, gauge changes at vital points required wagons to be unloaded and reloaded; incomplete tracks through the Taurus Mountains required goods to be transferred to wheeled vehicles then transported via treacherous mountain roads to the next rail and reloaded again. 'The supply of locomotives and rolling stock was entirely inadequate',[24] reported von Sanders. Roads for wheeled vehicles were rough while vehicles were underserviced and

---

[23] Von Sanders, L., *Five years In Turkey*, p. 7.

[24] Von Sanders, p. 28.

mechanically unreliable. Not only was it difficult to get their supplies, also the soldiers had difficulty going home on leave to see family and friends.

One stretch of their rail system that was effective, however, was the Hejaz Railway from Syria through Transjordan and the Arabian Peninsula. Built between 1900 and 1908 from Damascus to Medina at the order of Sultan Abdul Hamid II and made famous in the movie Lawrence of Arabia, the Hejaz Railway became a favourite target of allied airmen, Lawrence and his fellow Britons, the Arab fighters, and eventually the EEF and the Anzacs.

During the war, the Germans recognised this railway could move mines and submarine parts to intercept British shipping in the Red Sea and the Suez Canal. The Arabs, on the other hand, recognised the railway was an effective funnel for extended Turkish and German troop movements through the Hejaz and Arabian Peninsula into the troubled area of Yemen. This would impose even greater restrictions on Arab life and welfare with tax collections, military recruitment and loss of revenue from escorting pilgrims on their hajj to Mecca each year.

The Hejaz Railway, through the Arabian Peninsula, was the lifeline of Turkish supply to their 15,000-man garrison in Medina and the intermediate station at Ma'an with a 5,000-man garrison. It could have been cut and the garrisons captured but the last thing the British needed was 20,000 prisoners to feed and care for. Nor did they want 20,000 soldiers coming out to fight. So the strategy became: contain the forces in Medina and Ma'an and restrict their resupply along this railway. In the last few months it was totally cut and ended Turkish resupply.

Despite their difficulties, this rag-tag outfit did win at Gallipoli and Kut, crossed the Sinai twice. And, with German and Austrian advisers and munitions, they resisted the EEF advance for two and a half years.

When the Ottomans sided with the Central Powers, German trade and military missions remained and various military units and armaments quickly built up. These helped the Gallipoli campaign to Turkish success. By early 1916, the Germans provided scores of machine guns for a planned

Figure 2: A Hejaz Railway resupply train crossing a wadi that would have raging waters after rains (with permission from the Royal Hashemite Court of Jordan.

advance on Egypt; five anti-aircraft groups, heavy artillery guns, trench mortar companies, two field hospitals, transport companies, wireless operators and their 300th Air Squadron with new Rumpler aircraft. Later on, German special-forces infantry, additional aircrews and artillery joined the fray.

German officers General Liman von Sanders and Colonel Kress von Kressenstein would give outstanding service but much of their advice was ignored by the inexperienced super-egos of the Young Turks. Discord ran through the Turkish ranks so that, luckily for the Anzacs, their expertise was not fully used. Von Kressenstein proved himself an admirable adversary a couple of times, until Chauvel arrived on the scene and defeated him. Von Sanders was eternally frustrated as the supposed Commander of the Turkish Army while constantly interfered with by Djemal and Enver. Nevertheless, had it not been for German support in terms of capable officers, soldiers and their equipment, this campaign would have seen a much quicker end.

Desert Anzacs

This campaign was fought in the land of the Arabs where humans, mostly tribal, have occupied the Middle East for around 50,000 years. That's 45,000 before the first of the Egyptian pyramids. That's 49,400 before the Ottomans arrived. That's 49,800 before Governor Phillip and the First Fleet settled in Sydney Cove. And that's at the dawn of the global importance of oil.

In a tribal existence, populations came and went. Civilisations came and went. Nations never existed. Nomads wandered. Towns sprung up; some survived. Arabic Bedouin culture created over those millennia in that harsh environment provides hospitality to an unknown traveller for up to three days without question. That same culture allows five generations to seek retribution if a tribe or clan is dishonoured; that can be anything from a stolen goat or drinking from another's water well, to rape and murder; then another five generations for re-retribution; and so the circle of their life goes on. Tribal squabbles were endless. Today, tribalism continues throughout Arabic society, and is one of the many causes behind the current turmoil.

Figure 3: A modern Bedouin woman displaying cultural tribal tattoos, Syria.

The Good, the Bad and the Arabs

No empire has attained lasting control and many have tried. No race of people has dominated and many have tried. Arabs were not people of their own nations, or an Arab nation. They weren't motivated by national pride or imperial conquest and domination – they were simple folk who just wanted to live with their families, to raise their sheep, goats and camels and to live the simple life of hardship in their deserts or villages.

The *British Official History* of 1928 describes the Arabs this way:

> The Arabs, physically a fine race and highly intelligent, had always shown themselves politically frivolous, incapable of organisation and more prone to disintegrate than to coalesce.[25]

The Ottomans interrupted Arabic lifestyle. They had managed control of a sort for 430 years, setting a record for domination, and they were hated. They had subjugated the local Arab population, including tribal people as well as the independent cities and villages. None bore allegiance to one another, making domination somewhat easy.

The desert provides little and is a harsh landlord. One hundred years ago, the deserts of the Sinai and Arabian Peninsula were home to scavenging bands of nomadic Bedouin whose loyalty was firstly to their tribe and second to whoever paid them the most booty. They trusted no one; only payment in gold or weapons satisfied their loyalty and this loyalty shifted with each new enticement. They would scavenge the battlefield, plunder the dead and wounded then scurry away as troops arrived. New Zealand Trooper P.W. Burgess recorded, 'The Bedouin is the worst type of man I have ever seen. These devils have been known to come along at night time, dig up the dead, strip them of their clothing and leave them on the sand'.[26]

The young Anzacs Eric Boulton-Woods, Robert Bygott and Harry Wickham had no idea who or what they would find in

---

[25] MacMunn Lt Gen. Sir G. and Falls, Capt. C., *British Official History, Military Operations, Egypt and Palestine, Vol I*, p. 207.

[26] Woodfin, *Camp and Combat on the Sinai Palestine Front*, p. 20.

the Middle East. Nor could they have had any understanding of the influence they and their fellow soldiers would have there. No one had the foggiest thought of how events there would dominate today's world and all the global turmoil that explodes onto our TV screens and into our shocked and horrified hearts.

Figures 4/5: Today's more friendly Bedouin sharing their food and tea with our research group

5: The Anzac Journey

Optimism is the faith that leads to achievement.
– Helen Keller

Anzacs were intended to undergo training in England then be sent on to a European war. Fortuitously, Harry Chauvel was now in the Dominions Section of the Imperial General Staff of the War Office in London and he saw the lousy living huts being constructed for his troops. The approaching English winter reminded him that outdoor Aussies and Kiwis would not survive rain, sleet, frigid temperatures and poor construction. Chauvel appealed to an indifferent British staff that provided equally useless shelter for their own troops, to provide more sensible shelter for warm climate Anzacs. His appeal failed. Finally, engaging the support of Sir George Reid, the Australian High Commissioner, Reid persuaded Lord Kitchener to have our troops disembark in the warmth of Egypt for acclimatisation and training.

Trooper H.R. Williams recalls his first vision of Egypt from his troop ship:

> So this was the long talked of Egypt. It filled one with a kind of awe, as though the body of the dead and gone yesterday were laid out for view. It seemed to jeer at our most dearly prized achievements, firm in the knowledge that it would see the passing of the great nations of today, as of all that had been famous in the centuries gone by.[27]

They trained and acclimatised in the shadows of the pyramids and sphinx rather than the 'archipelagos of tents in knee-deep seas of mud'[28] that would have been their camp on

---

[27] Williams, H.R., *The Gallant Company*, p. 24.
[28] Bean, Charles, Australian War Correspondent from England.

the Salisbury plains of England. They were noticed as soon as they arrived:

> Suddenly Egypt was full of sunburnt strangers with glittering eyes and a total disregard for authority. They were blissfully naughty boys forever out on some gigantic spree. They raced each other up and down the Great Pyramids; within the first fortnight ten of them broke their necks falling down it. They sat all over the tops of trams, even the trams from wicked Em-Baba full of forty thieves; great laughing sprawls of Australians, disregarding the Egyptian tram-conductors, shouting, laughing, smoking, singing and constantly getting themselves electrocuted. They raced each other along the parapets of the Nile bridges for bets and fell off into the river and were drowned. They did everything no one else was allowed to do. In its uncounted years of history Egypt had probably never seen anything quite like them before.[29]

Anzac youngsters marvelled at the sights of men in long robes from the Arabian Nights, the pungent smells of open markets and open drains, the chords of strange musical instruments, the yells of honest and dishonest street merchants, the squalor of street urchins and the wails of the five times a day call to prayer from the Muslim minarets.

At times, they misbehaved. British Secretary to the High Commissioner, Sir Ronald Storrs, told of three lads who hailed an omnibus drawn by mules. The driver hastened his beasts and the occupants were seen to make unpleasant gestures towards the three. The three disconnected the mules then upturned the omnibus and its occupants and simply continued on their way over the horizon.

Old fashioned Aussie larrikinism notwithstanding, by landing them in Egypt's warmth, Australian officers demonstrated their natural care for their troops and morale. The troops trained in a mild winter climate similar to a North Queensland winter, thus avoiding the illness and disease so

---

[29] Napier, Miss P., *A Late Beginner*, p. 102.

prevalent in England. Chauvel found that the British Staff consistently demonstrated little understanding of Australians or the necessity for preserving the welfare and wellbeing of a fighting force. This deployment to Egypt was just the beginning of a campaign-long struggle with incompetent British staff officers.

A combination of youthful enthusiasm, lack of family supervision, regular pay, unlimited beer, group revelry and relief from non-combative boredom saw more than a few of the lads force their energies on the local population or brawl with the soldiery of other nations. The Anzacs played up whenever they were on leave. They were renowned for larrikinism, drunkenness, fighting, destroying property, contracting venereal disease, setting fire to or demolishing brothels, and failing to salute British officers. Yet they would chivalrously punish Arab men who dared mistreat their women or animals. Sometimes there was unforgivable ill-discipline. As Lance Corporal Joe Burgess recounts:

> Our chaps have built up for themselves a name in Cairo, they have taken it by storm and the tales they tell of it are very lurid. I believe that one chap of the First went so far as to pull the veil off an Egyptian lady's face – he did not last long afterwards, he was followed and his head was nigh severed off his body.[30]

Despite their mischief, they soon adapted, showing the world just how soldiering could be done. They were, and today's Anzac still is, highly regarded by all who share a space with them.

So just where did this new Anzac legend take place?

Australia's young soldiers may have heard of Palestine through school and Bible studies but few had any real idea where or what it was.

---

[30] Burgess, L/Cpl J., AWM MSS 1596/1.

Sinai, the wasteland peninsula wedged between Africa and Asia, is the gateway from Egypt to Palestine. Sinai has no potable water so treatment is mandatory. It's hot, very hot. Nothing worthwhile grows there. Civilised people didn't live there.

During the preceding millennia, armies and traders slid through Sinai's sands to get from one side to the other: the Romans, Alexander the Great, Napoleon and countless trade caravans. Richard the Lionheart avoided it. No one wanted it – a thoroughfare, that's all.

The northern top is splashed by the Mediterranean Sea with coastal swamps and sand dunes; a distance of around 150km before the firmer plains of Palestine. The western side has the Suez Canal in the north and the sparkling blue waters of the Red Sea further south, a scuba-diving paradise today. The eastern side is bordered by the Red Sea in the southeastern corner, framing an interior of mountains, sand and wild desert tracts stretching towards the Mediterranean. Sinai's interior is mountainous, boulder-strewn with raging waters in flooded wadis after winter rains, soft sand between rains, heat, flies and nasty creatures large and small. But this water disappears quickly, with only a few rocky cisterns retaining the brackish dregs. It shows no favour to man or beast who dare cross its network of mountains and wadis that run like veins in a sinewy arm. Only small bands of roaming Bedouin lived there full time.

Palestine, though, is a lot different. Palestine is the gateway from Asia to Europe. The Promised Land has desert and arid land, but it also has water and fertile pastures and, as the Jews have shown for hundreds, maybe thousands of years, it can be tamed and made into a rich resource. Over the centuries, Palestine's borders have been imagined, but never defined in a way that transient tribes or independent villagers can lay sole claim to, historically or today. It has never been a country; just the name of a region, although it's clearly a contentious name today where wars have been founded on emotion rather than any legitimate cause by non-Jews.[31] Its populations have changed, with no sustained tenure – some were villagers while others were nomadic and tribal. Armies passed in search of commerce, trade, conquest, religion, the custody of holy sites and the glory of kings; but none had permanent hold. Occupation seems to have given legitimacy to claims of ownership. But at this time, Ottoman occupation, reduced rainfall, deforestation, erosion and lazy citizens had seen Palestine deteriorate from Old Testament grandeur. 'Can these stony hills, these deserted valleys, be indeed the Land of Promise, the land flowing with milk and honey?'[32]

Jerusalem is the city at the heart of Palestine, the centre of three dominant world religions; Christianity, Judaism and Islam – they all share and recognise the same God. Sure, other theological issues divide them and until 100 years ago they shared the country. Life was generally peaceful. It contains the Church of the Holy Sepulchre where Jesus is said to have been interned and then resurrected; the Western Wall where Jews come to pray and leave prayer notes between the ancient stones of the Second Temple ruins; and the Dome of the Rock inside which is a stone from which Mohammed is said to have leapt on his horse into heaven.

Damascus, another historical city of major importance, was a city of beauty. Mohammed is reported to have ceased his journey on the hills above it and, astounded by its beauty, said 'I can only enter Heaven once'.

These were the lands of the Ottomans, shared with town and tribal Arabs, to be traversed by our young Anzacs over the next two and a half years, when legends and heroes would be formed. But where sons, fathers, brothers and mates would become involuntary sentinels, forever; in lands and among people they never truly understood.

---

[31] UN Resolution 181, dated November 1947, decreed a two State system, one Jewish and one Arab with Jerusalem being administered by an International Trustee. The Jews agreed but Arabs rejected the resolution and immediately went to war once the British Mandate finished in May 1948. Prior to this all religions and cultures had lived together in reasonable harmony.

[32] Dinning, Capt H., *Nile to Aleppo*, p. 60.

Figure 6: The stone in the Church of the Holy Sepulchre, Jerusalem, on which Jesus was reportedly laid after his crucifixion until his resurrection.

Figure 7: Dome of the Rock, Jerusalem, from where Mohammed reportedly began his journey to heaven.

The Anzac Journey

The war hadn't gone well for the British. Politicians, military commanders and the British public were rethinking the wisdom of this Great War. After Gallipoli, British strategy needed a major review.

Both sides had expected to achieve world domination by the first Christmas of the war. And from the first shot, this was like expecting gun control in America. Initial optimism was overwhelmed by under-preparedness, confusion, political interference, local corruption, indigenous rivalries, changes in military technologies and resultant tactics and sometimes, poor judgment. And so, Christmas came and went, as did the next one. And the next one. And the next one.

By early 1916, the Anzacs in the Middle East, a year and a half after war's booming start, faced global calamities. The year 1915 had been devastating for the Allies. France was in stalemate and slaughter, with opposing armies dug in their trenches. The Senussi tribes in western Egypt were in armed insurrection. In February, that Turkish force of 16,000 troops attacked the Suez Canal. In March 1915, a British and French naval attack in the Dardanelles, without air and ground force support, failed. In April 1915, the ground force invasion of Gallipoli took nine months before evacuation. In December 1915, a British army of around 10,000 men, mainly Indian troops with a half flight of the Australian Flying Corps, was besieged at Kut in Mesopotamia, to be wiped out or captured by April 1916. An allied force, not strong enough for offensive operations, was in Salonika attempting (with limited success) to sway Balkan nations to join the Allies. At the same time, the Muslim Arabs of the Hejaz in the Arabian Peninsula were seeking British support against their Turkish overlords. Jerusalem and Holy Land were still under Muslim control after 730 years. The Germans were not capitulating as predicted.

Turkish morale was stratospheric after Gallipoli and Kut. British morale was rock bottom after Gallipoli, Kut and the Western Front. The vacillating nations, tribal Arabs and Egypt's vast Muslim population were speculating that a German victory was in the offing and looking to side that way, while the Indian Muslims weren't sure. Thus, Great Britain faced the task of restoring domestic morale and international confidence.

The War Office could not decide on the best eastern strategy or time frame and its indecision confused their commanders and staff in Egypt and India. Finally, the War Office decided France was to be the priority. Troops in Egypt would be the empire's strategic reserve. The eastern theatre would become another sideshow to the main event on the Western Front.

Those wide-eyed Anzacs who'd left home full of enthusiasm were now tempered by the harshness of battle, the horrendous loss of their mates and the reality of the British Empire's inadequate military and political leadership.

So 1916 was a year of restructure. The survivors were back from Gallipoli and another Turkish attack against the canal was expected. The War Office and political leaders eventually realised they had to set some serious strategy for the east. Although the Western Front remained the priority and the troops in Egypt remained their strategic reserve, the Suez Canal had to be defended and the Turkish threat contained. A re-think and change of emphasis resulted in a new commander with a more 'active defence'. General Maxwell went back to England.

On 9 January, 1916, General Sir Archibald Murray became the Commander of the newly titled Egyptian Expeditionary Force. To know the Turks were coming was no masterstroke of deduction. To know what to do about it was. Murray had a keen sense of big picture strategy. But his understanding of battlefield tactics and use of troops proved deficient.

He drove the principle of active defence, although confusion reigned about Turkish numbers and intentions. Murray projected that 250,000 enemy troops could be launched against Egypt. The War Office, paranoid about France, stated there were more like 100,000 to 150,000. Thus, the allocation of Allied troops between France and Egypt was a bone of contention. Murray became alarmed. His Gallipoli-experienced infantry divisions were withdrawn to France and his force dwindled. He was left with undertrained and inexperienced Territorial soldiers from the farms and villages of rural, rain-filled England, the antithesis of desert. He did have the Anzac Mounted Division, on which he intuitively depended. By now the Anzacs were busting for a mobile, mounted fight after the septic trenches of Gallipoli.

Just as the British force in Egypt was dwindling, the Turks had their own problems. Some of their most battle-worthy divisions were also withdrawn from the Sinai Palestine region and sent to other areas of conflict in Mesopotamia, Persia, Greece and the Russian front in the Caucuses to placate German fears. And many troops were still in the Gallipoli area, unable to move because of the poor rail and road systems. The Germans became alarmed as the numbers and quality of British troop increased in France. They pressed the Turks even harder to attack the canal. Enver and Djemal needed little encouragement. Preparations continued with Colonel von Kressenstein taking charge.

General Murray had to stop them. Anticipating a Turkish attack through Sinai, Murray identified three possible routes of their advance on the Suez Canal.

A northern route was the shortest and most direct, from Gaza near the Mediterranean Sea in southern Palestine through El Arish at the edge of Sinai. It then ran along the sunburnt coastal sand dunes towards the hods, or watered palm trees, of Romani and Katia and finally on to the canal; a total distance around 120 miles.

A central route was used for the attack in February 1915, It went from Beersheba, 25 miles inland from Gaza, at the end of the Turkish railway in southern Palestine. Then, 65 miles of level but sandy and rocky ground southwest to Kossaima. It continued westward through jagged mountains and boulder-strewn wadi passes for another 85 miles to Jifjafa, then another 65 miles west to Ismailia on the canal. This route was devoid of water until winter rains had filled the rock cisterns around Jifjafa.

There was a southern route from Aqaba in the south on the Red Sea to the little village of Nekhl mid-way across Sinai then on to Suez. This was ruled out as too difficult as it was too long, too hot, too mountainous and waterless.

Murray's assessment was that a major Turkish approach would come from the northern route, with a second thrust through the central route. He believed his best strategy was to stop the Turks early on the narrow front at the start of those routes rather than wait for them to arrive on his wide canal doorstep – this made good sense.

His plan was simple and well founded – move troops along that northern route towards El Arish and stop the Turks at the gateway to Sinai. At the same time, neutralise the central route. A narrower defensive line would consume fewer men and resources. If only he could convince the War Office. To support the troops, he wanted to build a railway through the desert, build a pipeline to provide fresh Nile drinking water to men and animals, build a telegraph for communications and lay a wire mesh road for wheeled vehicles and foot infantry to march over.

Murray wrote to the War Office:

> Strategically therefore, the base of the defensive zone of Egypt against invasion from the east is not the eighty or ninety miles of the canal zone, but the forty-five miles between El Arish and Kossaima'.[33]

The War Office, perhaps recognising the lack of readiness of the British soldiery but failing to see the strategic sense of the proposal, gave limited approval. It restricted Murray's movement to Romani/Katia, a mere 25 miles from the canal and 70 miles short of El Arish. This gave an open invitation for the Turks to move unhindered on that precious water at Katia and close to the canal. Approval was given for the construction of the railway, pipeline, telegraph and mesh road. The War Office indicated it might approve later movement towards El Arish, but not yet.

---

[33] Keogh, Col E.G., *Suez to Aleppo*, p. 36.

Figure 8: A view from atop Mt Sinai; mountains ruled out the southern route.

It would be slow and arduous to move the army of foot soldiers, mounted troops, wheeled artillery vehicles, wheeled ambulances, engineering equipment, signals equipment and logistic support. It needed thousands of camels of the CTC to negotiate the northern Sinai's summer heat, soft sand dunes, fierce dusty winds, flies, scorpions, spiders and exotic bugs. This war of new technologies demanded huge supplies for aircraft, heavy artillery, armoured cars, vehicles for stores and transport, spare parts, fuel, signal and communication equipment, field hospitals, ammunition for rapid-fire weapons and water for man and animal. A light horse unit required three to four times the volume of water of an infantry unit. So the railway and pipeline had to be right on their heels.

Now to make it happen, information was needed. Information about the Turks, the Germans, the Arabs, the villagers and tribes, water wells, desert conditions, the religious effects of the mixed force, political matters. The list was endless. Then, it had to be assessed so strategic and tactical decisions could be made.

Desert Anzacs

One of the keys to success in war is to know the ground and the enemy: his strengths, dispositions, intentions and, what he is actually doing on the battlefield. Intelligence is critical. A lot of effort went into gathering information that had to be analysed and often cross-referenced. Similarly, to deny information, or provide misinformation, to an enemy is imperative.

Agents were planted in northern towns and cities to spy on the Turks; and vice versa, of course. Stories and tales were gathered from archaeologists and mapmakers, market vendors, from ships captains and crews, from wandering traders, from brothels and shopkeepers, from deserters and POWs. Local Bedouin would provide information for baksheesh, bribe money. It was often unreliable as they would say almost anything for money and the wilder their story the more money they expected. Confirmation was vital. Another aim was to win the hearts and minds of the local population and enemy soldiers.

In an innovation for airmen:

> True to the policy of avoiding all unnecessary harm to the natives, British aviators never dropped bombs on the town, but – they would unload packages of pamphlets, printed in Arabic, informing the natives they were being deceived; that the Allies were their only true friends.[34]

These encouraged deserters whose real value was their stories and information. 'The human substitute for providing up to date technical data was the growing number of Ottoman prisoners and defectors'.[35] This information seemed to come willingly, without much violence, and became the most reliable source of information. Many Arabic members of the Turkish Army surrendered and later even joined the Arab Army against their former Turkish colleagues. To stem the tide of desertions, Turkish officers shot Arabs they found in possession of these pamphlets.

---

[34] Aaronsohn, A., *With the Turks in Palestine*, p. 46.

[35] Sheffy, Y. *British Military Intelligence in Palestine Campaign*, p. 145.

Air warfare played an increasing role. Pilots would sometimes hand-draw sketches for immediate use by troops on the ground and to supplement inaccurate or non-existent maps. As time went on, significant aerial reconnaissance and photographs improved and added to the cross-referencing of information. And, what's taken for granted with today's technology, was in its infancy of wireless communication and eventually, quick transmissions added to the store of information. Unfortunately, the analysis of collected information was often ineffective.

Amongst all this General Murray had to choose which of his troops could make his active defence work: was it to be British farmers and shopkeepers or Anzac bush horsemen?

Desert Anzacs

6: A New Chapter to the Anzac Legend

The man who moves a mountain begins by carrying away small stones.
– Confucius

Another Turkish attack on the Suez Canal had to be prevented. This was the task Murray thrust onto the Australians in April 1916. He chose the Anzacs despite their refusal to salute British officers, their thirst for beer, and their inclination to burn 'gyppo' market stalls. In doing so, he sidestepped his under-prepared Territorial farmers, miners and shopkeepers. The next task, using British yeomanry, he'd regret.

Why he didn't stick with the Anzacs is a mystery, as

> Murray was deeply impressed by the resourcefulness of Anzac Mounted; they found their way surely about the trackless wastes of Sinai, carried out their tasks to the letter and returned to base in remarkably quick time. In fact, they were the only desert worthy troops at Murray's disposal in this early phase, apart from the few companies of the Imperial Camel Corps, which were themselves composed of Australians and New Zealanders.[36]

Murray had been told by the War Office not to go beyond Katia/Romani. He had to be there first to control that water and he also had to destroy the central route to deny its water to the Turks. Thus, concurrent action on the two routes was planned. He knew that only the Australian horsemen could find their way in the wilds of the Sinai sands and mountains. His British and Indian troops were not yet trained or acclimatised to desert warfare. His first action was to send a reconnaissance patrol to Wadi Muksheib on 21 March. Captain A.E. Wearne of 8th ALH

---

[36] Hill, p. 70.

Regiment (an experienced scout and Boer War veteran) led a patrol of 15 officers and 95 soldiers to determine the extent of Turkish water procurement works and the viability of this route for their future advance on the Suez Canal. The horsemen were a bit disappointed as there was no fighting, so no test of light horse tactics. But, several key lessons came from it that soon proved invaluable to all the mounted Anzacs.

The first was that the Australian horse could move faster in sand than the camel, surprising everyone but the horse. Moreover, the Australian horse didn't need as much water as was first thought and its endurance, continually proved in later operations, exceeded expectations.

In a new concept, the Royal Flying Corps (RFC) airmen preceded the column to advise on enemy and conditions ahead, allowing the column to move faster than it could when sending mounted scouts ahead. This brought a significant increase in speed to the mounted troops and began the fruitful cooperation and admiration between airmen and horsemen.

Finally, wireless was used between the column and HQ to communicate more effectively than pigeons and riders, increasing speed of information.

The air patrol detected Turkish water-drilling activity and troop movement near Jifjafa, further into the central route. Murray needed that water destroyed and the route nullified. He gave this task to the Australians.

So Chauvel tasked the 3rd ALH Brigade commanded by Brigadier J.M. Antill to destroy those wells. This would be the first significant operation conducted by Australians in the Sinai campaign. Critically, it would be Australian commanded. It was therefore the first real test of Australian tactical ability and leadership.

Antill tasked the 9th ALH Regiment under temporary command of the experienced Major William Scott (commissioned in 1903 and a Gallipoli veteran) to conduct this operation. The Australians didn't have all the support troops they would need for a mission of this type so Scott was given an untried and multinational force of mixed religions, cultures, languages, degree of training and commitment to join his eager troopers. His force comprised:

- HQ: Major Scott in command, with Captain Wearne (from the earlier patrol to this area); Captain Macaulay, Royal Artillery (he'd been with Wearne on the previous patrol) as Arabic speaker and Intel Officer; Captain Ayris of the 3rd ALH Brigade (with fifteen years' experience in the British Army).
- 3rd ALH Squadron of seven officers and 122 other ranks. 3rd ALH Field Ambulance for medical services with wheeled ambulances. Engineer detachment (British) for cistern destruction. Wireless detachment (British) for communication services. RFC detachment for liaison with No. 14 Squadron RFC.
- The CTC group of one officer, 97 native drivers and 195 camels for equipment transport. The Bikaner Camel Corps (Indian) of one officer, 24 other ranks for escorts. A detachment of 29 troopers of the 9th ALH Regiment as additional escorts that the British staff ordered should be dismounted for some unexplained reason. In addition, one squadron of the 10th ALH Regiment at Serapeum and aircraft of No. 14 Squadron RFC were on standby to assist.

Scott's force comprised 18 officers, 303 other ranks, 175 horses and 261 camels. They had to carry all their food, water, weapons, ammunition, personal gear, demolition and communication equipment, medical needs and facilities for prisoners. The mission given Scott was to verify reports of a Turkish force around Jifjafa, destroy water wells, report other water along the route, capture enemy personnel and identify other works and defences in the area.[37]

To integrate that lot into an effective fighting force would be an immense challenge for Major Scott. To add to the immensity of this challenge, intelligence informed Scott that the enemy's strength scattered nearby was over 1,000.

---

[37] Underwood, Col J., *Raid on Jifjafa*, Part 1. He quotes AWM4 Item 10/3/5 War Diary: 3rd ALH Bde April 1916.

Managing five or six nationalities with six or eight languages might not be everyone's dream job so Scott's hurdle would be the man-management of his diverse force. This had to be a success. It was Australia's virgin operation; everything was on trial and everyone knew it.

Realising the stakes, Scott selected the lightest soldiers and the fittest horses from his own regiment. Scott also had Hindus, Sikhs and Muslims of the Bikaner Indians, untried in battle; and the class-conscious British officers of engineers, signals and the RFC, who had no comprehension they must accept the military orders of a colonial officer, and the detested 'gyppos' as camel drivers. Food requirements differed. Their religious requirements varied (Muslims stopped to pray five times a day). Six languages required interpreters and who knew what could get lost in translation at critical times in battle? Very few officers or soldiers of any nationality had previous battle experience. Co-operation of all these groups was essential, compounding Scott's challenge even further.

His next momentous consideration was the logistics: one that is often the difference between success and failure and so often overlooked by scholars. Screw this up and the operation is dead in its tracks. A major factor was how to keep his men and animals hydrated throughout the mission. The camels carried some water in fantasses (five gallon bags secured on the side of the camel) but this would not be enough for both the outward and return journey. The only known wells en route were twenty miles southwest of Jifjafa and would be essential to refresh the horses on the return; that water had to be secured.

Water, forage, rations, ammunition, explosives and various stores had to be carried by the pack camels of the CTC with their walking drivers, escorted by the Bikaners mounted on their camels in their normal manner. But the 29 horse deprived troopers of the 9th ALH Regiment were here on foot. A British staff officer, in allocating troops and resources, decreed they could ride spare camels had they wished, but couldn't have saddles. This may have made sense to someone.

Desert Anzacs

Wheeled ambulances and medical staff with their lifesaving equipment were horse drawn, but wheels and hooves sank into the sucking sands. Some time after this operation, wheels were fitted with wide rims called pedrails to counter their tendency to sink into the sand.

This young Australian officer was soon to be tested big time. The troopers knew they too were on trial, their skills on display. They were proud and straining to go; but planning and preparation had to be completed, rehearsals conducted, orders given and letters sent home.

Departure day was 10 April. Early on such a day is an occasion for personal reflection; life is not eternal. Soldiers are people who have human needs and feelings. They complete final packing, conduct animal and gear checks, weapon checks. Soldiers and horses tuck into good food. Letters are written home. They reread the last letters from loved ones, say a prayer or two and go over the final orders: get to know the other nationalities and rest.

Just after last light, the slow-moving CTC group with its Bikaner and walking ALH escorts, slunk into the desert below a canopy of flickering stars; they needed no other light. They'd arranged for the fighting mounted group to meet up with them around midnight the following night.

Next morning the RFC patrol buzzed the area to identify any unfriendly Bedouin or Turkish patrols. Today, all was clear. This combined use of aircraft with ground forces was successfully unfolding after the earlier patrol by Captain Wearne.

> The association between the two services continued throughout the campaign and brought about a striking increase in the speed and effectiveness of mounted troops. Cavalry forces with airmen as advance-guards were able to travel swiftly and without fear over unknown enemy country instead of being compelled to probe their way slowly mile by mile, as in the days before flying.[38]

---

[38] Gullett, *Official History of Australia, Vol VII, Sinai Palestine*, p. 70.

A New Chapter to the Anzac Legend

The mounted column took off just after lunch into the summer heat; man and beast festooned with the black fog of flies that harassed eyelashes, burrowed into nostrils, clambered over parched lips and vibrated in ear canals, infuriating all except the flies. The column walked 50 minutes, then rested and watered for ten minutes, making a one-hour cycle. Hooves, feet and wheels sank into the sand. The horse-drawn medevac carts had difficulty keeping pace. However, when the ground eventually hardened, they caught up. Moving into the night, guided by glittering stars and with their outback experience transferred to the dunes, the mounted group caught up with the waddling camels and their escorts around 11pm. Night had brought relief from the furnace and flies.

Soon after the 5.45 sunrise, the morning RFC patrol spotted scattered camps of Bedouin known to be Turk friendly. A pilot dropped a note to Scott at 7.15; by 7.30 his troops were gone. Scott sent an ALH Troop to scout around known water cisterns, friendly Bedouin having reported a Turkish patrol in the area some days earlier. The troop gave the all clear and the force moved up to the cisterns where the horses gulped their first water after 22 hours of desert marching. Watered and rested, Scott sent out three patrols to scout for new water sources and signs of the enemy; tracks were located but not their makers. Nine new cisterns were located and although some were empty, the winter rains had dumped an estimated 140,000 gallons of good water into others: a gift from Allah for a future attacking Turkish force.

A forward base was established on a plateau where the Bikaners and dismounted light horsemen with the CTC remained with the wireless detachment, all under the command of Captain B. Ragless. The attack group moved out at 7pm, having dined on culinary marvels of cold bully beef from the tin and biscuits, washed down with water; no fires for a warm meal or cuppa for fear of being spotted by restless Bedouin. A night move was needed to enable an early morning attack, intended to shock the Turks. This also avoided the furnace and flies of the daytime but made moving ahead slow – seems camels have poor night vision and bump into things. A march of 40 minutes earned 20 minutes' rest in the hourly cycle. It took seven and a half hours to cover twelve miles through the most rugged terrain so far. The guides earned their pay that night and the Australians once more demonstrated their outback skill of movement through unknown country.

The march continued through those last hours before the sky lifts its darkened shutter, slowly letting in the first rays of light that glowed upon jagged, saw-like mountains. Then, the sun cascaded across the dunes to hit unprotected eyes and bring back swarms of sticky flies.

On the morning of the attack, 13 April, men and horses muffled into position. As the time approached a strong westerly wind blew up that billowed red dust as thick as a London fog. It prevented observation of the positioning of the attacking force by the Bedouin but it hampered clear observation of the battle zone by Scott and his sub-unit commanders.

Attack or wait, pondered Scott.

Attack. Launched at 9am the horsemen galloped into the Turkish camp, the Turks taken completely by surprise. The Anzac conquest of Sinai Palestine had catapulted into a successful start.

Two hours later, Corporal S.F. Monaghan, the only friendly casualty, and six dead Turks were buried. Wounded Turks were treated and 34 prisoners and documents were secured; the Royal Engineers destroyed the cisterns denying the use of the central route to any future Turkish force.

An Austrian water engineer was captured with two key documents. One stated that the German 300th Squadron, equipped with fourteen Rumpler aircraft, would soon arrive in Beersheba; these were a more advanced aircraft in speed and firepower than those operated by the Royal and Australian Flying Corps (AFC).

The second document was a Turkish report of the known water resources in the Sinai and Turkish progress on drilling and extraction:

This document was to prove of great value for operational planning in the coming months of the Sinai campaign as it indicated those areas where water was available in sufficient quantities for mounted formations to operate.[39]

The success of the operation was summed up by one of the ALH lads when he wrote home:

> Last week I went out into the desert on a bit of a jaunt after Turks. We were away five days and covered 60 miles before striking their camp. They ran for their lives but they didn't stand a chance, for our fellows ran them down with their horses, and captured the lot. We feel very proud of ourselves, as we were the first ALH to go into action as a unit on real Light Horse work.[40]

One of the founders of Qantas, Paul McGinness, was a light horse sergeant on this mission and was awarded the Distinguished Conduct Medal (DCM) for conspicuous gallantry. He would later become a pilot in the AFC, gain a Distinguished Flying Cross and ace status, before going on to Qantas.

Major Scott had demonstrated what a mixed force of nationalities, cultures, religions and customs could achieve with good leadership, respect for any ethnic group, discipline and planning. The Australians had justified General Murray's confidence in these troublesome colonials. However, Scott still had to get his force home with his prisoners and those prized documents.

Finally back in camp, the troops were rewarded with food, water, maybe a beer, a shower, clean clothes and sleep. At 3pm, advice was received from General Murray that Major Scott had been awarded a Distinguished Service Order (DSO) and Sergeant P.J. McGinness and Corporal P. Teesdale a DCM. Subsequently, a MID was awarded to captains T.C. Macauly (Royal Field Artillery) and A.E. Wearne, Corporal J.T. Talbot and Lance Corporal J.W.J. Miles of the ALH.

---

[39] Underwood, Col J. 'Raid on Jifjafa Part 1'. *Sabretache*, Vol XLII, Sep 2001.

[40] Duguid, C. *The Desert Trail: With the Light Horse through Sinai to Palestine*, p. 76.

Desert Anzacs

The award of a DSO to a Commander is the customary manner of recognising an exemplary performance by the whole unit when not all deserved soldiers can be given an individual award.

With the Australian-led force heading towards Jifjafa, Murray's attention turned to the northern route, where the railway was already under construction and heading east into the desert. However, he wasn't aware a Turkish force of 3,500 men had arrived.

At the same time he'd sent the Australians to the central route and victory, Murray thrust the northern route at his Zone Commander, Lieutenant General the Hon. H.A. Lawrence. Lawrence was a citizen soldier who had won no credit in battle. He'd been a junior officer in the Boer War, passed over for promotion then retired. He became a banker and Member of Parliament. At the outbreak of the Great War, he was recalled as a junior officer (rank of major) following ten years of banking and parliamentary service. He rose unbelievably from major to brigadier-general within eleven months of staff work and no field, command or battle experience. How he then passed through major general to lieutenant general within another year is unimaginable, but likely based on his political connections.

Lawrence threw the unprepared 5th Yeomanry Brigade at that northern route. There were two objectives for the yeomanry: to protect the pipeline water for the use of the railway's engineers and construction crews; and prevent the Turks who may advance towards the canal from using what was now the only natural water in the Sinai at Romani/Katia. The citizen yeomanry, led by officers of the landed gentry with virtually no battle experience, were not up to a desert fight. They weren't fully trained or acclimatised and they weren't supported with heavy weapons. Gullett notes:

> The wealthy young men of England, when they respond whole-heartedly to the nation's call to arms, tend to treat their newly acquired military responsibilities in a very sporting manner. They do not in the least mind

dying for England, but they like to go to war casually and, if possible, in comfort. They ask that the wretched business shall not, except as a last resort, too seriously alter their regular habits of life.[41]

Regrettably, when they die though, they take their soldiery to the heavens with them.

The Turks were battle hardened. They were desert savvy. Unhindered by British patrols or aerial reconnaissance their presence was unknown; bloody nuisance that 3,500 Turkish troops led by Colonel von Kressenstein popped up. Their mission was to confirm that the railway, reported by friendly Bedouin, was under construction and the suitability of water for their major force to approach the canal.

Further, the Turks well knew the British dispositions. German aircraft had flown over regularly, photographing and mapmaking; even bombing camps on 20 and 21 April. Experienced commanders would have seen this as a signal that something was afoot and would have increased patrolling and vigilance. But not the inexperienced Lawrence and his yeomanry.

Lawrence dispatched the mounted 5th Yeomanry Brigade to Romani, under the command of Brigadier E.A. Wiggins. Wiggins was sent without artillery support (gun carriages couldn't move easily through the soft sand), inadequate machine gun protection and accompanied by unmounted soldiers of infantry and engineers who struggled through the soft sand. On arrival, Wiggins split up his meagre force as wind flutters loose paper. He scattered detachments to camps at Dueidar, Oghratina, Romani and Katia, on little bumps in the sand dunes beside the railway or into the desert where only a palm tree or two gave shelter from the sun. Brackish water had the horses turn up their noses and the troops couldn't drink. Wiggins and his HQ staff then trotted off (with Lawrence's permission) towards Bir el Mageibra twelve miles away, in pursuit of a Bedouin reported enemy force. He lost all communication with the rest of his brigade.

---

[41] Gullett, p. 81.

Desert Anzacs

The scattered detachments, each 100-300 troops strong, were separated by soft sand six to twelve miles apart. At a travel pace of one to three miles per hour, it was impossible to support one another or receive assistance from reserve troops, had there been any. Such methods totally ignored rudimentary soldiering. The result:

> In the tragic engagement which followed, the folly that first sent the brigade alone into the desert, and afterwards which divided it into isolated camps, ignorant of the enemy's movements, was redeemed by the magnificent fight to the death carried on by the slender force of yeomanry officers and men.[42]

The morning oozed thick fog. The air was still, no smells drifted on a breeze to warn of Turks who usually stank of stale BO and bad food. The yeomanry sent out their early patrols. All returned before the fog lifted then failed to conform the protocol of leaving sentries outside their perimeter. This again displayed their lack of training, discipline and common sense. They had no idea a silent enemy in large number was upon them.

The Turks had divided their 3,500-man force among the scattered British detachments. Under cover of the morning fog, they attacked. The outgunned Tommies fought hard and well while ammunition lasted, inflicting heavy casualties. The yeomanry had horses; they could have escaped. But they had foot infantry and engineers in their midst, proving it was foolish to mix mounted and foot soldiery. Yet, as revealed in British reports, the cavalry commander courageously stayed to support his ground troops. Moreover, they didn't have artillery or adequate machine gun support. They could not contact their brigade commander. Many were wiped out, one group at a time while many others surrendered once they ran out of ammunition.

The fear of young men, from both sides, is highlighted by one of the survivors, Private William Joy of the 5th Battalion The Buffs, East Kent Regiment:

> I heard the high-pitched sound of a machine gun, I

---
[42] Gullett, p. 85.

A New Chapter to the Anzac Legend

> buried my face in the sand, too paralysed to think, all my manhood oozed away, I was terror stricken like a little child.[43]

Wounded and left on the field that day, he returned to his unit the following month. He would be killed in action in February the next year.

Wiggins had no idea the Turks were in his area until after the first shots. From the time he set off on his quest for the imaginary enemy force, he had no communication with his scattered detachments. Command functions did not exist. Junior commanders had no one to call. With 600 of his soldiers dead, wounded, captured or left on the battlefield for the Bedouin to plunder, Wiggins and his HQ, not having found the reported Bedouin and with no one left to command, withdrew.

Colonel Keogh in his training directive summarises the follies as follows:

> Brig-Gen Wiggins arrangements for the defence of the area were basically unsound. Instead of dispersing his force in penny packets beyond effective supporting distance of each other, he should have established a firm base from which he should have patrolled the outer areas. Not one of his detachments was strong enough to offer effective resistance if heavily attacked. The crowning error however, was the move of the Brigadier with his Brigade HQ to Mageibra. Consequently, when the Turks attacked he was unable to exercise any control over his Brigade. By such actions a formation commander may display commendable personal courage, but he shows little aptitude for his real duty.[44]

The greater failure was Lawrence. He failed his senior command function by allowing Wiggins to scatter his troops piecemeal, then to wander off with his HQ. He mixed mounted and foot soldiers in a terrain of soft sand. He failed to provide artillery and machine gun support. But the politically connected general kept his job.

---

[43] Woodfin, E., p. 65.

[44] Keogh, Col E., p. 42.

Desert Anzacs

The Turks were slapping backs and celebrating. They had completed a successful raid. With no intention of holding the wells at Katia or Romani on this occasion, they withdrew. On their return to Jerusalem with their British prisoners paraded, they achieved enviable propaganda and celebration. They had succeeded beyond expectations, aided by Lawrence and Wiggins. 'Coming so soon after Gallipoli, the success upon Sinai was of great political and moral value to the enemy'.[45] British morale took another hit – offset by Anzacs and Major Scott's success. In a mastery of understatement the British Official History total summary of the action goes like this:

> The affair at Katia was a lamentable occurrence, resulting as it did in the total loss of three and a half squadrons of Yeomanry. Otherwise, it had no effect, except to delay the progress of the railway for a few days.[46]

The British Official History doesn't mention the cause of this disaster. And of-course, there is no mention of the affect these unnecessary deaths had on their families at home.

Murray, once more acknowledging the skill of his Anzacs, belatedly sent the Anzac Mounted Division towards Romani to attempt a rescue of the 5th Yeomanry and to secure the area. Chauvel had ordered the 2nd ALH Brigade, commanded by Brigadier Granville de Laune Ryrie, to move as quickly as possible. Ryrie was probably the biggest man in the Anzac Mounted Division (AMD) and one of the most capable horsemen and rifle shots in the Anzac force. Born and raised in the bush, he had shown himself to be a true soldier. Possessing courage, leadership, an understanding of mounted tactics and professionalism, he was adored by his men but he had no tolerance for those who chose convenience over proper soldiering.

After a nightlong march through undulating sand dunes with a canopy of stars for dim light, the 2nd arrived all too late. They moved in and occupied the remnants of British camps. They were appalled at what they found among the bodies of

---

[45] Gullett, p. 88.
[46] MacMunn and Falls, p. 169.

soldiers and animals: the signs of the comfortable officers' lives with no military significance – golf clubs, white tablecloths and silver candelabra, sherry decanters and folding chairs. The stigma of this consistently poor professionalism among the self-professed professionals who demanded salutes, impacted on the Anzacs throughout the whole campaign.

However, the Turks were gone. But they would be back with the goal of blocking the Suez Canal and tying up the largest British force they could. Only the Bedouin were to be seen, desecrating the bodies of Britons and Turks to relieve them of jewellery, clothing, boots, weapons, food, money and whatever glittered. They slunk off into the desert as Ryrie's troopers approached. In a heart-wrenching footnote to this, a few months later a party of British soldiers with a chaplain and a doctor, Major O. Teichman, came upon

> dead horses, camels and men; Turkish and British bodies lay in the open, most stripped naked by Bedouins. They had been buried by the Australians but the desert winds and shifting sands had pushed them to the surface again. The sad little party had come to record the graves and bury their comrades, stacking sandbags on top of them to protect the new gravesites.[47]

The day after Ryrie's brigade arrived General Murray placed command of the Romani/Katia area to Chauvel. The road from Palestine through Romani is

> the oldest road in the world – that great highway connecting Africa with Asia, followed from time immemorial by invaders, Egyptians and Babylonians, Crusaders and Saracens, and Napoleon. The Christ Child fleeing with his parents from the wrath of King Herod came down this road to Egypt.[48]

Now it was the task of Chauvel and his mounted troops to prevent the Turks following the path of the Christ child to Egypt. The only major water along this road that the Turks could use was the brackish goo at Romani and Katia a few

---

[47] Woodfin, pp. 13-14.
[48] Powles, Lt Col. G., *New Zealanders in Sinai and Palestine*, p. 15.

miles east. Reading from the correct military tactics handbook, Chauvel concentrated the AMD around Romani and ordered extensive patrols in strength with his officers and young soldiers trained and ready, prepared to fight; bayonets, not golf clubs; machine guns, not sherry decanters. Patrols were not to engage in protracted fire fights. They were to maintain their mobility, report Turkish movements and seek water for future use, utilising Major Scott's captured document. And aerial reconnaissance would soon provide advance intelligence for Chauvel. The Anzac presence was growing and Major Scott had shown what the Anzacs could do. However, this area was still under the operational command of General Lawrence who, through the grace of God, accepted Chauvel's proposals.

Colonel von Kressenstein had scored another victory over the British. But things were about to change. This would be von Kressenstein's last victory. Chauvel was now on the spot and had replaced shepherds with warriors.

The Anzacs were in town.

Really, who were these abrasive and disrespectful ruffians that didn't salute British officers? Not even soldiers a few months ago; what did they know about soldiering compared with the centuries of tradition that British regiments had built up?

7: The Anzac Warriors

'Lewis' (Sgt Yells) was an Australian, long, thin and sinuous, his supple body lounging in unmilitary curves. His hard face and predatory nose set off the peculiarly Australian air of reckless willingness and capacity to do something very soon.
– Captain T.E. Lawrence (of Arabia)

The Australian Light Horse (ALH)

They were horsemen first and always; they thrived outdoors where mate looked after mate, a man's word was his bond, and his horse was his cherished possession. Chauvel's official biographer, A.J. Hill notes, 'a man's first and last care, in camp, at sea, training in Egypt was his horse'.[49]

The real horseman would tend his horse, lay up his saddle and clean his rifle before thinking of feeding himself. In a hostile desert, special attention to the horses was vital. Fleas, ticks, mange, fly infestations had to be controlled. Reduced water and poor quality feed all meant extra brushing, grooming and often hand feeding. Captain Robert Ellwood of the 2nd ALH Regiment illustrates this attitude:

> Our horses were the first thing we considered the moment we got up and before we went to bed. They were the whole of our life and they repaid us.[50]

Mateship and loyalty are sometimes nebulous concepts, hard to define. A definitive aspect of it though, was unwritten. From their introduction in April 1916 until the end of the war in October 1918, the light horsemen would not allow a sound mate to be captured. This approach was largely successful, as

---

[49] Hill, p. 48.
[50] Ellwood. R. *War Memories of Robert Ellwood*, www.jcu.edu.au/ellwood.

after two and a half years only 73 light horsemen were taken prisoner, most wounded. Not one officer was captured. Yet in achieving this remarkable record there was much sacrifice, another circumstance of men at war. But mateship is mateship and one had to try to save a mate; that's all there was to it.

Neither cavalry nor mounted infantry, light horsemen were mounted riflemen, 'adept at scouting and screening, or any operation where speed and flexibility, rather than firepower, were essential.[51] They rode to battle, dismounted and advanced like infantry. Their mobility gave them the power to inject or extract themselves with speed and surprise. Rifle and bayonet were their weapons, supported by machine guns, artillery and their mates in the air.

They were mostly country boys. They were born there, grew up there and enlisted there. Troops and squadrons were enlisted in towns to form regiments within their regions. The regiments formed into brigades and the brigades into the Australian Mounted Division that in the beginning looked like this:

- 1st Brigade: regiments from New South Wales, Queensland and South Australia/Tasmania, commanded by Brigadier Charles Cox.
- 2nd Brigade: two regiments from New South Wales and one from Queensland, commanded by Brigadier Granville Ryrie.
- 3rd Brigade: regiments from Victoria, South Australia and Western Australia, commanded by Brigadier John Antill.

Chauvel believed his initial make up of officers and men gave him a huge advantage. All the COs and a few of the officers had seen service in South Africa; many commanded pre-war light horse regiments and around 85 per cent of the men were from the bush and country towns. They could ride, shoot and live in the outdoors without too much supervision – it was seldom necessary to tell them what to do. Chauvel,

---

[51] Hill, p. 36.

however, was a stickler for discipline and training – practices that permeated throughout the whole of the Australian force, leading to their superior successes.

There was more to a light horse brigade than the fighting horsemen. Each brigade usually had:

- *A field ambulance section.* Unarmed stretcher-bearers, drivers and medics.
- *A sanitation section* for field hygiene and disease management.
- *A veterinary section* for the many animals.
- *A machine gun squadron.*
- *A signal troop.*
- *An engineer field troop* for water acquisition and purification.

What's a horseman without a horse? The waler was the favoured horse of the Australians. Called a waler for one of two reasons; some think it was because they came from New South Wales. The more likely reason is because that's where they were mustered to from around Australia and sold to the Army. Normally they carried 280 pounds and required six gallons of water a day; on operations they would often carry up to 500 pounds and go without water for 48 to 60 hours.

Completely overlooked in the telling of military history is the devotion and contribution of the members of the Australian Army Veterinary Corps (AAVC) and the animals they cared for. Michael Tyquin tells of

> the enormous difficulties faced by the AAVC in maintaining the health of the animals that were so critical to the prosecution of the war. Mobility of warfare was the defining feature of the Middle Eastern theatre until late 1918.[52]

Fighting mobility had to be maintained by the combatant horses and camels and the extraordinary number of baggage

---

[52] Tyquin, M., Forgotten Men: *The Australian Army Veterinary Corps*, p. 143.

camels, mules and donkeys provided by the ELC and CTC or there wouldn't be a war. These animals had to be in top fitness by maintaining their health and stamina, feeding and watering them, preventing sickness, treating wounds and sores. Making the veterinarians' job even trickier was that horses and camels don't talk – they don't tell what is wrong, unlike a soldier at a first aid post.

Each brigade had a mobile veterinary section (MVS), absolutely vital to the animals on the battlefield. The MVS worked cooperatively with the Veterinary Hospital and Australian Remount Units (ARU) in Egypt.

Replacement horses came from those ARUs. From March 1916, there were two squadrons in the ARU, one commanded by Captain H. Reid and the other by Australia's famous poet and journalist Captain A.B. (Banjo) Paterson. Each squadron had 196 men and around half were rough riders whose task was, where necessary, to break horses and make them rideable to go to the front line.

> The function of the ARU was to take the horses which were shipped to Egypt from many parts of the world and prepare them for issue in the shortest possible time. The riding horses had to be trained to ride in troop formation, to lead, and to accept the improvised tethering of breast, ground and heel ropes which were used in the desert. No mount was issued until it was steady enough for a fully equipped soldier to get on and off under any circumstance.[53]

Life as a rough rider was no easier than that of a light horseman being shot at. All manner of horses would arrive for training and there was considerable urgency to get horses forward as the poor buggers were constantly being shot or blown up. And breaking horses is not easy work. It took a special man to do it. Many of these roughies were specialist exhibition riders from circuses, horse-breaking arenas, stockmen, buckjump and carnival riders. Roughies like Dick

---

[53] Kent, D., *The Australian Remount Unit in Egypt, 1915–19, a Footnote to History*, undated.

Bell from Victoria and Roy Standbridge from Queensland were joined by Stan Breneger and Ernie Green from Dungalear near Walgett, with Jack Carter from Goonoo Goonoo and Jack Dempsey, who was the leading rough rider in Australia at the time – not household names today, but, in those days, they were well known in the country circuit of Australia.

These guys weren't typical soldiers. Army uniforms weren't their go. They wore mufti outfits that suited their trade rather than attempting to look soldierly spick and span. They were a rough lot that got the job done.

Around 160,000 Australian walers were sent to war. Regrettably, only one came home for cost and quarantine reasons. But now controversy arises. The healthy ones were to be handed over to local 'gyppos' to be used for domestic chores.

It is well known through soldiers' diaries, letters and texts that the light horsemen cherished their walers that had carried and been loyal to them for years. It has been reported that some soldiers were so heartbroken to think their horse would be given to a local and be mistreated, as was the local habit and custom, that they contemplated shooting their horse. This would have been illegal and a court martial offence. Regrettably it seems, quite a few of these treasured companions stepped into a wombat hole and, out of kindness, had to be put down. Those soldiers were never charged.

New Zealand Mounted Rifles (NZMR)

New Zealand horses and horsemen were as revered in 1916 as they are today. Win a Melbourne Cup, win a desert scratch race, or charge into battle. They were as rugged a bunch of horsemen and shooters from their colonial surroundings as the Australian horsemen. The NZMR Brigade, commanded by their own Brigadier Edward Chaytor, included the Auckland, the Wellington and the Canterbury regiments, each made up of squadrons from home towns and villages such as the 3rd

Auckland, the 4th Waikato and 11th North Auckland squadrons. The Brigade had a similar structure to an ALH Brigade.

The ALH Brigades and the NZMR Brigade became the Anzac Mounted Division.

The Imperial Camel Corps (ICC)

Of all the animals that Noah released from the Ark, the camel would have to be the most puzzling. It is distinguished by knobbly knees, a long neck that puts its head right in the rider's line of sight, that head with teeth like the front grill of an early Holden, a lump in the middle of its back where you're supposed to sit, a get up and down speed to match finger nail growth, but it does have good-natured eyes; could such a beast go to war? History says this beast is a proven warrior. Mounted Persians routed the Lydian hordes; Romans were scattered by camel-mounted Arabs in North Africa; Napoleon used them in Egypt 100 years earlier; British forces rode to the relief of Khartoum on them 30 years earlier; and they were imported to the Australian outback initially for the Bourke and Wills expedition in 1840 for their endurance and mobility.

Figure 9: Monument to the Imperial Camel Corps in the Embankment Park beside the River Thames, London.

The camel is special; it is splayfooted, which allows its pad to spread over sand and not sink like a horse's hoof; it has four stomachs for water retention with a five or six-day holding capacity; its long eyelashes protect its eyes from blowing sand and persistent flies; it can close its nostrils to exclude dust and flies; it can carry a soldier, his equipment and six days rations with water plus food for itself; it can cover 30 to 40 miles a day in desert sands; it is easily hobbled; but it stinks, spits, snarls and provides the riding comfort of a springless cart over rocks. Yet it was about to become the secret weapon of the British Army and the Anzacs.

In January 1916, cameleers were initially selected from the Australian infantry. Despondent at being whipped at Gallipoli and at the loss of many of their mates, now they were being forced to march, drill, salute, train needlessly (to their thinking) in the heat, sand, dust, wind, flies and squalor near Cairo. To escape this military madness volunteers scrambled to the Camels, praying for relief from that boredom and for the excitement that had caused them to enlist.

The cameleers gained the reputation of being more like bushrangers than the average digger. Major Oliver Hogue (pen name Trooper Bluegum), a camel company commander wrote:

> We were quite as careless as the average soldier, and no more moral. So we were always losing saddles, or fantasses, or dhurra bags, or brushes, or head-ropes. And we had to make good the deficiencies. But we never thieved – at least not within our own company lines. True, it was no great crime to commandeer a saddle or camel from headquarters.[54]

But not just anyone can handle a camel effectively. According to all records these rough Anzacs took a fraction of the time to train that their British counterparts needed; even local Arabs were amazed. The Australian bushman was a natural at animal welfare and health management. Lieutenant Colonel George Langley, a camel battalion commander, states that from the commencement of training, 'four weeks later the first company marched out as trained cameleers – a record in quick training.

---

[54] Hogue, Major O. (penname Trooper Bluegum), *The Cameliers*, p. 18.

The usual period of training given to British troops was five months'.[55]

Mind you, they weren't pets. Trooper John Robertson: 'When a camel attacks a man he uses his teeth first, then attempts to crush the life out of him by kneeling on him and pounding him with his bony knees.'[56]

The early structure of the ICC was:

- 1st Anzac Battalion: Nos. 1, 2, 3 and 4 companies (Australian)
- 2nd Imperial Battalion: Nos. 5, 6, 7, 8, 9 and 10 companies (British)
- 3rd Anzac Battalion: Nos. 11, 12 and 14 companies (Australian) and No. 15 Company (NZ)
- 4th Anzac Battalion: Nos. 13, 17 and 18 companies (Australian) and No. 16 Company (NZ)
- Artillery of the Hong Kong and Singapore Mountain Battery known as the Bing Boys (so named after a London East End play but the reason seems lost in time)
- 26th Camel Sqn Machine Gun Corps (Scottish Horse)

The total strength of this initial brigade was 2,800 all ranks with 1,800 riflemen. Twelve of the 18 camel companies were Anzacs.

The camel became a means of transport for the infantry soldier. These ungainly brutes were totally conspicuous and never able to hide their approach from the enemy. They were so gangly when it came to lowering themselves for the rider to mount or dismount, that speed of military action was just not a happening thing. Hence, camel soldiers would ride some distance from their objective and dismount then footslog to fight infantry style.

[55] Langley and Langley, *Sand, Sweat and Camels*, p. 47.
[56] Robertson, J., Cameleers and Camels at War, www.nzhistory.net.nz, quoted by Philip Walker in *The Jaddah Diary of Captain Thomas Goodchild*, *T.E. Lawrence Society Journal*, Vol XXI (2011/12), No. 2, p. 48.

The Australian Flying Corps (AFC)

Before the Great War, air warfare was thought about in Australia, Britain, France and Germany with mixed vision. Some prepared for military aviation while Britain's top soldier, General Sir William Nicholson, Chief of the Imperial General Staff, visioned in 1910; 'The aeroplane is a useless and expensive fad, advocated by a few whose ideas are unworthy of attention'. France's top soldier, General Foch, followed this in 1914 with 'The aircraft is all very well for sport – for the army it is useless.'[57] War quickly changed that unimaginative thinking.

Germany was at the forefront of air technology, their aircraft attaining higher speed, greater climb and better armaments. France and Britain followed. At the outbreak of war Britain had 113 aircraft in military service, the French Aviation Service 160, the German Air Service 246, Turkey four. Hard to know why, but none of them were armed and the first pilots used sporting pistols or rifles to pot shot at one another. Production flourished and during the war France produced around 68,000 aircraft with 52,000 of them lost in battle.

Far from backward though, Australia had recognised the potential of air warfare in 1909. The government offered a hefty prize of £5,000 to develop a flying machine for military purposes. It was never awarded despite worldwide interest. From 1912, our government planned, then early in 1914 created the Central Flying School (CFS) in a cow paddock at Point Cook in Victoria.

The rapid pace of technological innovation was matched by a rapid change in creative uses. Key tasks that aircraft could perform were discovered, experimented with, and refined: observation and reconnaissance, photography, tactical and strategic bombing, ground attack, aerial combat, naval and artillery fire direction, air-to-ground communications. With the growing importance and influence of aircraft came the need to control the skies and thus the fighter plane was born.

---

[57] Isaacs, K., *Military Aircraft of Australia 1909–1918*, p. 42.

At home, in early in 1914, trainee pilots were called for. The rush of thrill-seekers was overwhelming but only four were selected. The CFS held its first class with instructors Lieutenant Eric Harrison, an Australian, and Lieutenant Henry Petre, an Englishman. Their first pupils were Captain Tom White (a 26 year-old part-time officer and metalworker) and Lieutenants Richard 'Dickie' Williams (a regular army officer), George Merz (a part-time officer and medical graduate) and David Manwell (a part-time officer and commissioning agent). Some of the unchosen scampered to England where they signed up for the RFC and the Royal Naval Air Service (RNAS).

Over the next two years, more courses were conducted. More than 3,700 men eventually served in the AFC, with hundreds more racing to England and the RFC and RNAS. In fact, almost half the RFC and RNAS comprised soldiers from the dominions of Australia, New Zealand, Canada, South Africa and India.

No. 1 Squadron AFC

On 16 March 1916, No. 1 Squadron, under the command of Lieutenant Colonel Edgar Reynolds, departed Australia on the *SS Orsova* for Egypt and glory. Twenty-eight officers and 195 soldiers, with rifle and bayonet for the soldiers, a sword and pistol for the officers (why airmen would want a sword remains an official mystery), a few motor vehicles, a bunch of tools and stores, but no aircraft.

From its underwhelming start in the cow paddock, Australia provided the only dominion air arm to the war effort. They arrived in Egypt on 12 April, around the time that Australia's Major Scott was leading the first Anzac operation to success at Jifjafa and the Turks were annihilating the British Yeomanry Brigade. On arrival after a month at sea, No. 1 Squadron was as surprised as the AIF and British HQ that they had not been expected. Confusion arose: what to do with No. 1 Squadron AFC?

The answer came a week later. The three Flights of No. 1 Squadron were attached to the RFC Squadron for further training. Although they were named No. 1 Squadron AFC, the British system couldn't tell the difference between their No. 1 Squadron in France and the Australians in Egypt, so they called it No. 67 Squadron RFC. Nevertheless, throughout this book I refer to it as No. 1 Squadron AFC, which the British finally recognised in April 1918.

Over the next six weeks, the squadron mechanics, riggers, instrument fitters, armourers, drivers and telegraphy specialists underwent further training with their British counterparts. All the officer observers and five of the least experienced pilots were beamed into an English summer for combat training. The remaining pilots worked with their British squadrons and had British observers allocated to them.

The demand for pilots in the coming months and years was such that volunteers were called for in theatre. Many a light horsemen worn out by the bitter, windswept sand of the deserts, raced to take up the opportunity to engage in the air, scrambling to get into those flimsy aircraft. As it turned out, the Australian horseman proved to be a skilled pilot or observer. A quick eye, steady hand, mental agility and initiative, willingness of heart and individuality of purpose were part and parcel of being an Australian bushman. And they understood mounted tactics, could tell the difference between an advancing or retreating enemy, and so they worked well with the mounted forces. Saddles were swapped for cockpits, bayonets for bombs and rifles for cameras and into the heavens they flew.

Of course, an airfield and an air force consist of far more than the derring-do pilots and observers upon whom the glory and honours are most often cast. The squadron had three large tent hangars. Each would house six aircraft per flight. Flimsy, flapped around by wind, they needed care. Each floppy tent was an Aladdin's Cave that housed:

- An armourer responsible for the eighty guns of the squadron, eight types of ammunition, bombs and their fuses.

Desert Anzacs

- A photographic section to produce much-needed maps.
- A self-contained mobile workshop when called out to a downed aircraft.
- An instrument fitter workshop with clocks, watches, speed indicators, altimeters, tachos, cameras and bombsights.
- A wireless telegraphy section provided air-to-air, air-to-ground and air-to-ship services.
- A stores section kept everything from bully beef to spare trucks and uniforms to aircraft wings.

The Australian fliers stood out to the British senior officers, just as their brothers in the light horse and ICC were doing. The RFC Commander, Major General Geoffrey Salmond, a respected British officer, officially recorded 'the rapid training and mobilisation of the squadron reflected great credit on the

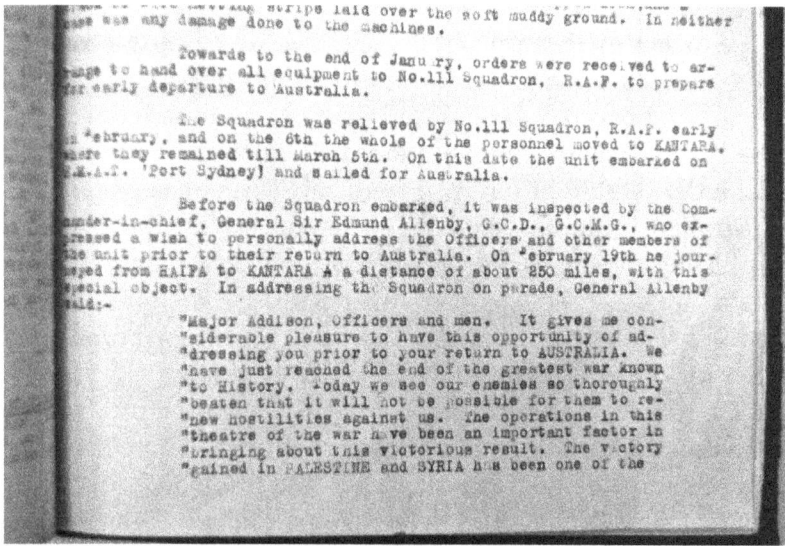

Figure 10: War diary of No 1 Squadron AFC, the message from Commander-in-Chief, General Sir Edmund Allenby (1).

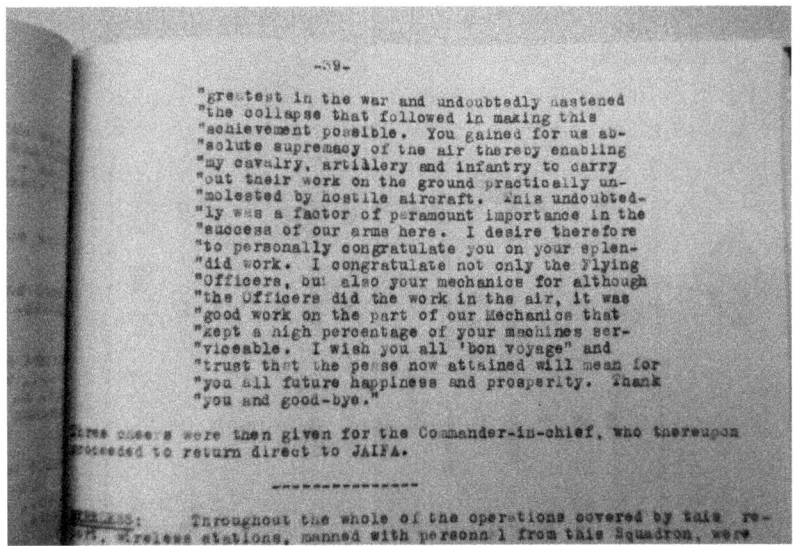

Figure 11: War diary of No 1 Squadron AFC, the message from Commander-in-Chief, General Sir Edmund Allenby (2).

industry, keenness and discipline of officers and all ranks'.[58]

In the early years, the AFC and RFC aircraft in this theatre were constantly at a huge disadvantage to the Germans in terms of speed, climb, manoeuvrability, endurance and weaponry. However, the pilots proved to be significantly more industrious, adventurous and daring than their German counterparts and this somewhat made up for what they lacked in equipment. It was only in the last nine months of the war that our airmen finally received superior aircraft that made the skies their own.

Aircraft and airmen of both sides quickly became indispensable. Their courage and pioneering acts were the origin of the most effective but frightening form of modern warfare that is still developing in the 21st century. For the first time in battle, aircraft bombed and strafed often without warning, terrifying the unsuspecting ground troops and their animals. This new horror quickly impacted the minds of the

---

[58] Molkentin, M., *Fire In The Sky*, p. 58.

young ground soldiers and civil populations. Psychological warfare was being redefined.

One of the most prominent Australian flyers was Lieutenant Dickie Williams. He had been one of the first four graduates from the CFS. Having joined the army at the age of nineteen he was commissioned into the South Australian Infantry Regiment in 1911. And a damn serious fellow he was; didn't smoke, drink, gamble or cus – how he survived the army is anyone's guess. But, better than survive, he flourished. Wowser or not, Dickie was esteemed by the men who flew with him and he led them to stardom.

As an early graduate he was in the first group to sail for Egypt and fly among the clouds above the holy lands. On arrival in Egypt, he was appointed flight commander with the rank of captain. By January 1917 he was promoted major and given command of No. 1 Squadron, the only Australian squadron in Egypt. And he flew fearlessly. On one occasion he landed behind enemy lines to pick up a downed mate; on another, he attacked Turkish ground troops with the aircraft machine gun while under heavy anti-aircraft fire, scattering those troops. He eventually became the Commander of 40th Palestine Wing RAF, which included five British and the Australian squadrons. He was awarded a DSO and several MIDs. It was well known that Dickie, promotions, honours and all, was strict, formal and correct but fair with his crews; they respected and almost loved him. This man would be a career flier and achiever like Harry Chauvel of the light horse.

Meanwhile, enemy eyes were still focused on the Suez Canal.

Part Two

Sinai to Gaza

8: Survival in Sinai

The greater the difficulty, the greater the glory.
– Cicero

As it happened, von Kressenstein waited three months for reinforcements and resupplies before attempting another advance on the Suez Canal. The Anzacs waited at Romani, the War Office having decreed advance no further.

They patrolled, trained and became desert hardy but the youngsters got restless in the empty expanse of the desert. Rest periods were spent writing sacks of letters for the censor, rereading letters from home, playing cards or the never too far-away two-up game, writing stories for unit newsletters, drawing cartoons, rereading newsletters, writing diaries, relating or inventing stories to better the last man's. Some had cameras for posterity.

They extracted enormous ticks, lice and fleas from their horses and camels that had become their best friends, and snaffled enriched feed to keep them in peak condition. They treated their cuts and sores to avoid infection and downtime. The men spent extra time grooming their best friends, trimming lengthening hair, cleaning fly deposits from eyes before infection set in, then stroking ears and noses in comfort; only then attending their own needs. The baggage camels and donkeys had to be maintained to keep up their stamina too.

Desert life was a drag on health, stamina and the mind. To maintain sanity, time off was taken when possible with trips to the Mediterranean or back to the Nile for a swim for men and horses. Creative boredom, yet quiet optimism and cheerful confidence, filled their days; but they craved action.

Desert Anzacs

By now, the whole concept of warfare had changed. They already understood the horrors of cavalry and artillery, but there were new tactics to consider now. Bombs and machine gun fire, seemingly coming from nowhere, were dropped from aircraft and caught men and horses in the open with no anti-aircraft protection. No ground commanders, on either side, had experience with aircraft and even pilots had little idea what they should be doing. Both sides experimented. Chauvel and von Kressenstein were to learn new battle procedures, unimagined moments before. These two opposing commanders demonstrated that some old dogs could learn new tricks. All the while, the ground warriors were shattered with fright from sudden attack. Captain Frank Weir described it as, 'Bombed again – it's just like a hawk in a chicken yard the way we scatter when the Hun flies over us.'[59]

In that desert, April to June is springtime. After winter rain, flowers in their green, yellow, lilac and reds, fluttered like flags in the breeze. Grass sprang up to create a green carpet on which horses and camels grazed with excitement and temporary relief from the slate blue and brown-red rocks that quickly returned. Such colourful glory brought soldiers to a halt to witness this break in brown monotony. Nostrils, clogged with dirt and dust, rejoiced at the fresh smells of sage and rosemary. Then summer came on quicker than promises at a political rally. Temperatures hit 125^0 F (over 50^0 C) in the shade. Out of Egypt came the khamsin, the wind that delivered dust storms for hours or days, during which seeing an extended hand was near impossible, and breathing through filter cloths mandatory. But it gave respite from bombing and snipers:

> Terrific sand storm. Think the worst we have had. One gets very tired of the eternal sand, in face, ears, tucker, bed etc. One continual dust storm for a week – We must eat lbs of it cooking in the open – but mud is worse.[60]

---

[59] Weir, Capt F., diary, MLMSS 1024, Mitchell Library.
[60] Weir, diary MLMSS 1024.

Survival in Sinai

Figure 12: A recent desert dust storm, *khamsin*, of the type endured by Anzac soldiers throughout this campaign (courtesy Ms Rana Naber).

Romani soon filled up with the AMD, the 52nd British Lowland Infantry Division, the 42nd British Infantry Division, artillery and companies of cameleers. Night and day, all suffered ferocious months of desert life before the next Turkish assault.

Meanwhile, this period enabled the Anzacs to hone their desert skills. They often moved at night to avoid the heat, flies and detection by the enemy watching for columns of dust. Navigating the featureless dunes where the one in front looked the same as the one behind, using just starlight and a compass became second nature to them, much the same as in outback Australia. Long rides, with men and horses suffering water deprivation while skirmishing with local Bedouin and Turkish patrols, became critical experiences that gave the Anzacs what would prove to be their advantage in the years ahead.

When a beachgoer shakes his towel and sand flicks into unsuspecting eyes, it causes momentary discomfort. But suffering this desert every day for weeks, or months, is another matter entirely. As Trooper Robert Bygott explains after one of his skirmishes:

> And laying in the scorching sun made a fellow terribly thirsty and the water in your bottle was hot, but better than nothing and you had to be very careful as you didn't know where the next one was coming from. It was terribly hot on this ridge and your rifle seemed red hot after a few rounds and the sand blowing in the magazine caused it to jam.[61]

It's simple physiology: if you don't drink, you die. If you don't drink enough, you still die. The initial water allocation to troops in the desert by British staff, who could be accused of more care about water for their scotch each evening than troops in the desert received, was 'in the hot sun only one water bottle a day to drink and wash with', as Bygott continued. Moreover, each man carried just a one-quart bottle or just over a litre. This would never be enough to maintain health and fighting fitness. After one extended patrol,

> some of the horsemen found themselves walking back into camp after their horses collapsed. The stragglers wandered into camp late that night and the next day 15 men (including four officers) went to the hospital for sunstroke and heat exhaustion.[62]

It took Chauvel's appeal to General Murray himself to force the staff to provide a second water bottle. This common sense decision could easily have been made at a lower level in an effective HQ.

In another case, on 6 August as they chased the retreating Turks, the British 125th Infantry Brigade of 2,802 men, recruited from the chilly lands of northern England, started a summer desert march. At the end of that day, only 2,162 remained upright – over 600 had fallen out from heat and exhaustion.[63] They too had only one water bottle each.

[61] Bygott diary.
[62] Woodfin, p. 45.
[63] Woodfin, p. 45.

Figure 13: Watering the horses, Palestine, 9 Jan 1918 (courtesy, Australian War Memorial).

With nearly 1,000 horses in an ALH brigade, huge supplies of water were needed quickly. Isolated desert wells weren't sufficient and sand collapses choked the cumbersome pumps provided by the British supply system. Lieutenant Colonel Wilson commanding 5th ALH Regiment, a Queensland-based unit, suggested using the spearpoint pump that he had used in the dried creeks and riverbeds of the Ayr district in North Queensland. This was a 2.5 inch tube that pierced the sand easily. Connected to a small pump, it could produce good volumes of water within minutes, filling water troughs and bottles. The British admin staff saw no merit in the time saving, volume producing, spearpoint pump. This just exasperated the contempt the Anzacs held for them and gave weight to the English philosopher John Locke's comment: 'New opinions are always suspected, and usually opposed, without any reason other than they are not already common'. Officers gathered unit monies to buy the required materials and the engineers made hundreds of pumps for Anzac troops. It was no surprise that once the British cavalry got into the desert and saw the merit of the pump, they convinced their British staff to provide them. Horses adapted to the brackish water fairly quickly but for humans, it had distressing effects, so field ambulances and engineers provided a form of purification, even though the taste took some getting used to.

If you really want to tick off good soldiers, give them rotten food. Not that the admin staff didn't buy good food. They just neglected to have it packaged and stored properly. By the time good meat and vegetables arrived, it was often flyblown and putrid. Long past was the view that enduring hardship was a soldier's contribution to a war effort. Morale suffers when the problem is unremitting. As reported by Colonel Keogh:

> The rigid and unimaginative attitude of the administrative staffs created amongst the troops out on the hard, blazing desert a feeling of resentment and bitterness towards the people who worked and lived in comparative luxury in Egypt.[64]

Sometimes it wasn't the unimaginative staff. '[A] Bedouin dog stole our bacon this morning, that's worse than a Taube'.[65]

Now, some thought this preparatory life in the sand was not the real war. For two years, death and destruction had ignited France. The desert itself wasn't the issue; many thought that a man should be fighting in that real war where fellow Australians were already dying and others needed help. To be retained in the desert on minor actions was a slight to their manhood and neglect of their mates. In fact, the Australian public at home thought the desert war was a picnic compared with the real fighting in France and those dirty, wretched trenches.

It was reported that a female volunteer who made socks for the men abroad placed a note in one pair saying, 'I hope this goes to some Australian hero in France, but not to any of the

---

[64] Keogh, p. 47.

[65] Weir, Capt F., diary.

cold-footed Light Horse'.[66] Well, those socks were received by a badly wounded light horseman evacuated to the 1st Australian General Hospital (AGH) in Egypt. There was a disastrous consequence to his emotional state. The Australian public had not been told and so had no idea what desert warfare was really like.

Feelings were strong among the Anzac soldiery. Some felt they would be better utilised fighting with their brothers in France. Corporal Francis Curran and six of his mates from the 7th ALH Regiment, reflecting that war in the sand was something less than war in the trenches, stowed away on a ship bound for Marseilles in the south of France, intending to make a worthwhile contribution to the war that they'd signed up for. They'd already served at Gallipoli so knew what fighting was like. Curran had been awarded a DCM and MID for service there, so he was no slouch. Lieutenant Colonel George Langley reports:

> With several others, he was caught at Marseilles and returned, crestfallen, to Egypt. Technically, he had deserted and when the facts about him were found out he was placed under arrest.[67]

As more and more felt this way, discontent brewed among the Anzacs. Lack of action, the monotony of constant patrolling and training, the barren desert and its not so barren creatures had the men believing they should be in France, where the real war was. Chauvel and his officers had a tough time maintaining discipline and purpose among the youthfully exuberant Anzacs. The Turks and the Arabs were soon to help out and Corporal Curran wasn't done yet.

---

[66] Hogue, p. 47.
[67] Langley and Langley, p. 46.

Desert Anzacs

9: The Arab Revolt

Eagles don't flock
– Ross Perot

On 10 June 1916, nearly two years into the Great War, Sharif Hussein bin Ali, Emir of Mecca, sat alone by the window in his room in Mecca, rifle in arms. He had the weight of an uncertain future on his shoulders. His thoughts were troubled – king or conquered? Should he rise against the Muslim Ottomans, his cousins in religion but who'd denied him an Arab kingdom in their empire; or side with the Christian British, who'd offered to support his dream of being the king of an Arab nation?

Sharif Hussein fired that rifle and started what became the Arab Revolt:

> In the beginning of the Revolt, the Sharifs would be leading poorly equipped Bedouin tribesmen unsupported by artillery or machine guns against trained, combat-experienced, and well-led Ottoman soldiers with artillery, machine guns and airplanes.[68]

There was no going back. Now they needed British weapons, money and food. How did it get to this? Why would fragmented and quarrelsome tribes take up arms against their overlords of 400 years?

After the Young Turks took control in Constantinople in 1909, Arabs who had previously enjoyed positions of influence were removed. In 1915 and 1916 in Beirut and Damascus, around 30 Arab intellectuals and notables were executed by public hanging. Government services were centralised and throughout the empire, local officials no longer had influence or

---
[68] El-Edroos, *The Hashemite Arab Army*, 1980, p. 58.

The Arab Revolt

received pay. The Hejaz Railway had taken away the livelihood of Arab tribes who had previously provided escort, food and transport services to Hajj pilgrims. It also gave Turkish troops movement to control the Arabs even more. The Young Turks planned to assassinate Hussein. He found out and was not happy. Taxes on Arabs and conscription increased. Services reduced. The Arabs had little representation in the empire. Hussein had been invited by some but not all Arabs to lead them in revolt and form an Arab nation. The British offered help.

But creating unity among Arabs, even to create an Arab nation, was like rounding up eagles: they just don't flock. The Damascenes in Syria and the Hashemites in the north of today's Saudi Arabia were ready to revolt; the Rashids in southern Arabia but far away from the action were on-side with the Ottomans; the Idrissi and Sauds were so far away they didn't care and didn't support Hussein. Individual tribal leaders made it clear to Hussein they would not leave their tribal lands to engage but if Hussein brought an enemy to them and gave them gold and weapons they would fight within their boundaries. Britain provided weapons, ammunition and food while Australia gave the gold sovereigns to meet their needs.

However, history shows that Arab unity disintegrates quicker than morning fog.

To Hussein's delight, however, the Arab Army quickly captured Mecca, accompanied by wild jubilation that Arabs excel at. Within a couple of days, a troupe of tribesmen descended upon the Red Sea port town of Jeddah adjacent to Mecca, to be held back temporarily by the Turks. Ships of the British Red Sea Patrol (RSP) bombarded the city while aircraft of the RNAS from HMS Ben-my-Chree bombed and strafed. The city surrendered to more wildly cheering tribesmen. Other Red Sea cities capitulated. But despondency soon hit at Medina; its garrison of 15,000 Turks held out. Very soon, the Arab Army, without artillery or machine guns, and with only antiquated rifles, came to a standstill.

Figure 14: Wadi Rumm where Arab forces with British advisers camped and prepared for attacks on Aqaba and the Hejaz Railway.

In October 1916, the junior Lieutenant T.E. Lawrence, together with Sir Ronald Storrs, Oriental Secretary and a British intelligence officer in the Arab Bureau in Cairo, arrived in the Hejaz on an investigative mission to recommend how much more support Britain might give. Lawrence met Emir Feisal and the two formed a lasting bond, as later did Feisal with Lieutenant Colonel Pierce Joyce, who would eventually command the Hejaz mission and became Lawrence's boss. Storrs recommended extensive support to the Arab Revolt that the British eventually committed, including Anzac support. General Murray, doubting Arab capacities and aware of his commitments to the Western Front gave limited support of weapons, food and money but greater use of naval ships and aircraft. Although these had an initial morale-boosting effect to bring tribal warriors to the revolt, there was insufficient support for a prolonged campaign and pretty soon the revolt was suspended inland where naval and air bombardments couldn't reach.

The Arab Revolt

Now, we ought to put the legend of Lawrence of Arabia in its proper historical context. The highly lauded but often questioned Lieutenant (later Lieutenant Colonel) T.E. Lawrence, became a liaison officer to Emir Feisal. He was part of a group of around 50 British officers that assisted Hussein, his sons and their cause and did a remarkable job in harnessing independent Arabs into a force of fighters, eliminating the need for an all-British force. In fact,

> these negotiations introduced a gallant band of Englishmen who afterwards played so influential and picturesque a part in the Arab war against the Turks. The young Englishman shows nowhere to such advantage against men of all other races as in the handling of coloured alien peoples. The personal hold gained in wild Arabia by a few British civilians and officers, unsupported by military force, was a rare tribute to their courage, tact and ability.[69]

Lawrence became the most famous, thanks only to brilliant marketing by the American showman Lowell Thomas after the war, his own book *Seven Pillars of Wisdom* which is the only English text written of this part of the war, and the epic 1962 movie *Lawrence of Arabia*. His name evokes the mystery of the east, flowing Arab regalia, horsemen and tribal treasure-seekers. Famous in British circles, yes, but,

> Lawrence is nothing. There were many British officers and it is not right to say one man did all those things to make the Arab Revolt succeed. That film says Auda Abu Tay only wanted gold, it is an insult to the Bedouin.[70]

The film is a great Hollywood epic but gives a false impression of what really happened. For one, the real Lawrence was 5 foot, 5 inches tall, compared with the six foot something actor Peter O'Toole. Similarly, Lawrence's own book *Seven Pillars of Wisdom* is a classic piece of literature; semi-fictional

---

[69] Gullett, p. 75.

[70] Author's interview with Muaffaq Hazza, a Jordanian historian and archaeologist, on 28 July 2015.

literature. Lawrence himself says it recounts his memory of his part in it, not an accurate history of the Arab Revolt. Yet the world took it as an accurate historical record. Parts of the book are correct but his role in some events is vastly over-emphasised, to the exclusion of other British, Australian and Arab individuals, who played as much or more an influential part. General Chauvel, when asked by the Australian War Memorial to advise whether it should purchase a copy of *Seven Pillars*, wrote a thirteen page letter highlighting many of the deficiencies but, in the end, recommended it as an investment in literature rather than a record of history. It does give an insight and draws attention to a vital part of the Great War that could otherwise have remained totally neglected.

Lawrence is to be acknowledged but not lauded, having done no more and perhaps less militarily than other British officers. He did, however, wield quite some political influence. It should also be understood that Lawrence, for the rest of his life, felt the guilt of what he said was British betrayal of the Arabs. He refused an investiture by the King and joined the Royal Air Force (RAF) as an airman and then the tank corps as a private to hide from the world. He assisted the Arabs wherever possible, but his efforts were shattered by fractured Arab leadership and by British, French and American politicians.

The real effect of the Arab revolt wasn't felt militarily until a year after its commencement when the Arab forces captured Aqaba at the head of the Gulf of Aqaba. Critically, the Arabs did not side with the Ottomans, or life may have been a lot worse for the EEF and the Anzacs. Some historians and academics argue that the military consequences of the Arab Revolt were insignificant; perhaps so. This, however, displays no appreciation for the effect of the revolt. Effects often result from what you don't get as much as what you do.

Had the Arabs sided with the Ottomans, something like this might have resulted: Sharif Hussein could have followed the *jihad* call then attracted millions of Muslims in Egypt and India. Hundreds of thousands of their Muslim soldiers

The Arab Revolt

and labourers could have denied their services to Britain or deserted and taken up arms against the British. All Arab forces could thus have opposed the British. This would have denied to the British, armed support, intelligence, local food supplies, drinkable water for men and animals, safe rest areas for troops on leave, guides and local knowledge. Similarly, the Sultan's jihad would have had a much greater impact as all Arabs may have joined, rather than have it dwindle as it did.

In terms of its impact on this region, Aqaba[71] would not have been captured, denying Britain the use of its seaport as a supply route.. The Hejaz Railway would have given the Turks unrestricted logistic support of troops and materiel throughout the Arabian Peninsula and may have extended to the west to harass Allenby more effectively. The EEF right flank would have been exposed during its advance through Palestine. This exposure would have required tens of thousands of British troops, who weren't available, to defend it. The large Ottoman garrisons in Medina (15,000 troops) and Ma'an (5,000 troops) would have been available for redeployment against the British.

Similarly, the Muslim Egyptians may have risen against the British, denying them a stable logistic base, but also creating a guerrilla-style conflict and jeopardising British control of the Suez Canal. This would have denied the EEF the extremely valuable tens of thousands of workers in the Egyptian labour and transport corps, without whom the campaign would have failed.

Local tribes and villagers may have actively opposed and intercepted the British in Egypt, Sinai, Palestine, Syria and the Arabian Peninsula. Finally, the Gulf of Aqaba could have harboured German shipping and submarines against the Suez Canal and British shipping in the Red Sea.

One could argue that Hussein's shot saved the British Empire from a German victory in the Middle East that could have had a domino effect on Europe with a huge Turkish Army joining that fray. Had that happened, had the Germans

---

[71] The capture of Aqaba in July 1917 is described in a later chapter.

and Turks won the war, the rise of Hitler twenty years later wouldn't have been necessary.

Hussein's support for the British came at a price: he wanted to be declared 'King of the Arabs' and Caliph of the Muslims in a unified Arab nation with gold, weapons and food for his armies. If granted, this would have massive political ramifications in France and England, India and Egypt and, the whole of the Muslim world. Allied skullduggery and vague acknowledgements arose.

So the consequences of the Arab Revolt were significant and directly contributed to an eventual Entente success in the Middle East, and, arguably, the Western Front. Pre-war agreements forgotten, the Paris Peace Conference in 1919 and other political talkfests thereafter, divided the Arab lands of the Levant between France and Britain under mandates. Hussein bickered with Arab leaders, who also bickered among themselves, and failed to gain their support. British and French leaders titled him 'King of the Hejaz' only. He became obstinate and unyielding in post-war discussions that may have brought harmony. He even alienated his sons Feisal and Abdullah.

Today, there is widespread Arab discontent as a result of what they see as Britain and France's refusal to honour pre-revolt promises and a failure to declare a unified Arab nation. Arguably, that has contributed to the recent Arab Spring sparked in 2010 in Tunisia and spreading across the Arab world in 2011 along with the rise in global terrorism. But, in reality, Arab unity was and still seems impossible due to antagonistic tribalism, distrust and unresolvable sectarian beliefs that result in constant violence. An independent Arab nation was unlikely to manifest no matter what the Allies did. However, certain Arabs see the betrayal by the British and French as the sole cause of their plight today, failing to consider any of their own internal conflicts.

So what happened to Sharif Hussein? He did declare himself King of the Arabs and Caliph but, hardly surprising, failed to gain the support of all Arabs. Between 1920 and 1924

The Arab Revolt

he and his first son Ali lost battles with Abdul Aziz ibn Saud (both tribes supported the British during the war and both were now denied any assistance; sort your own issues was the mantra). He was exiled to Cyprus where he died in 1931. As a result, ibn Saud created Saudi Arabia that includes the Hejaz.

Emir Abdullah became Emir of Transjordan under a British mandate (then its first king, with Jordan's independence coming in May 1946 as the Hashemite Kingdom of Jordan). Emir Feisal was declared King of Syria at an Arab Congress in 1920 but the French kicked him out 100 days later so the British made him first King of Iraq. The French were given mandates over Syria (that became independent in April 1946) and Lebanon (that became independent in November 1943). Britain was given the mandates for Palestine and Mesopotamia. Britain's mandate over Palestine ended in May 1948 following United Nations Resolution 181 that recommended a two-state solution for Palestine (one Jewish and one Arab with Jerusalem to be administered by an international Trustee: Jews accepted, Arabs rejected). Once British troops withdrew, the State of Israel was declared and the Arabs immediately went to war against Israel. Today, the non-extremists of Palestinian leadership seek the two-state system that was rejected, while the extremists still seek the destruction of Israel.

Sharif Hussein bin Ali had set the world on a new path. And the war continued for the EEF and the Anzacs.

Meanwhile, back in Sinai …

10: Battle of Romani

A small group, well led, can change the world.
– adapted from Margaret Read

Three months after the loss of General Townsend's army in Mesopotamia and one month after the Arab Revolt started, back in the desert the folly of the fragmented dispersal of 5th Yeomanry hadn't got through to Generals Murray and Lawrence; their thinking was as mysterious as the occult. He now split the Anzacs.

In a letter dated 12 July 1916 Murray praised Chauvel:

> Whatever I ask you people to do is done without the slightest hesitation and with promptness and efficiency. I have the greatest admiration for all your command.[72]

You'd think an effective leader would want to keep such a force together, like you would a gold medal hockey team. No. Murray, even realising the Turks would soon approach Romani and Katia on their way to the canal, then divided these best troops into separate areas, like splitting your gold medalists into three teams. He scattered the Anzac Mounted Division so Chauvel had only two of his four brigades: the 1st and 2nd plus the Wellington Regiment of the NZMR Brigade. The NZMR Brigade (less the Wellington Regiment but with 5th ALH Regiment) went west under the command of General Lawrence, with an infantry division incapable of fast movement, as a reserve force. The 3rd Brigade went south under Murray's command with his reserve force of immobile infantry on the canal. But Chauvel was given what was left of the Territorial 5th Yeomanry Brigade and their untrained, inexperienced

---
[72] Hill, p. 74.

Battle of Romani

reinforcements. Made sense to no one but Murray and totally bewildered Chauvel.

To divide the Anzacs among British commanders who didn't understand light horse tactics and place them among immobile infantry, after acknowledging their outstanding combined work under Anzac commanders, was stupid. The suggestion by staff officers that the Anzacs could teach desert battle procedure to the inexperienced British troops instead of developing their own teamwork and battle procedures was offered, but is absurd. Yeomanry charge into the fight with swords; light horsemen approach then dismount and proceed as infantry with rifle and bayonet. Where's the synergy here? It's like putting those hockey players in a cricket team because they each have timber sticks to hit a ball. The early edict of the Australian Government that had placed our troops under the command of 'any' general the British chose had come back to bite.

Worse, Murray plonked himself and his reserve force at Ismailia on the south of the canal. Lawrence plonked himself and his reserve force at Kantara at the north of the canal. Both were so far behind any action and with difficult communications they were unable to observe any action to make sensible battle decisions. Nor could they quickly send those reserve troops to assist the battle when Chauvel called for them; which he did and they didn't.

On 19 July 1916, Brigadier Chaytor, Commander of NZMR Brigade, while observing from an aircraft of No. 14 Squadron RFC, detected around 8,000 Turkish soldiers with horses and camels, artillery and vehicles, moving towards Katia and Romani and therefore, the Suez Canal. Over the coming days airmen of No. 1 Squadron's B and C Flights, joined by No. 14 Squadron RFC, spotted further Turkish advances. Michael Molkentin reports:

> These (advances) abruptly stopped on 31 July, suggesting the Turks were in position and ready to strike ... On 1 August, Australian pilots Alfred Ellis and Lawrence

Desert Anzacs

Wackett joined a seven aircraft bombing raid on those positions. Richard Williams meanwhile tried his hand at directing fire from an offshore naval vessel on the Turkish positions ... It was a process that would come to play a vital role in British tactics, but was still in its infancy on this front.[73]

The attack was imminent. The scattering of the AMD remained.

For the battle scene, imagine a rectangle; on the east (or right-hand) side lying in a north–south direction was the British 52nd Infantry Division in trenches with fortified defences that were almost impossible to attack successfully and quickly.

Across the top in an east–west direction was the Mediterranean Sea; no problem. To the west or left side was the Suez Canal, with water and reserve forces, so the Turks weren't going to get there. At the bottom of the rectangle were the mobile horsemen with limited defensive works.

Chauvel wanted to draw the Turks into the sands where the mounted troops could savage them. His tactic was to allow them into the bottom of the rectangle where the 1st ALH Brigade would be waiting, lure them into the soft sands and heat where there was no water, tempt them by withdrawing his mounteds as if being defeated then attack with his mounted reserves and the infantry when the Turks were exhausted and dehydrated. To be effective, he would need quick access to the NZMR Brigade and his 3rd ALH Brigade and have the infantry available to move swiftly. Murray and Lawrence, with faulty communications when the landlines were cut, faltered. They retained the Anzac mounted troops and the infantry in their own reserves until the battle was nearly lost.

First, though, Chauvel (possibly in a state of disbelief but we don't know) must have wondered whether this could work with his mounteds divided. And why did he have two commanders anyway, especially these two; who was running this show? And, would the Turks really think that way?

---

[73] Molkentin, *Fire in the Sky*, p. 64.

Battle of Romani

On the night of 3 August, he found out:

> History scarcely presents an example of such complete conformity by an enemy taking the offensive to the plans and wishes of the defenders.[74]

The *British Official History* reports:

> The Turks were fully prepared to play the role allotted to them by the British command. The account given by Kress shows that it was exactly what had been anticipated.[75]

Around midnight the Turks, led by Germany's Colonel von Kressenstein, followed Chauvel's plan exactly. They bypassed the entrenched infantry and followed on the heels of a returning Australian patrol into their camp. Later, this was shown to be an error of Australian tactics by returning at the same time and the same way each night; that was never repeated.

The desert was quiet that night and starlight shimmered off the sands: moving objects, like people sneaking up, could be detected. As good training and a vigilant Australian sentry system would have it, the Turks were spotted and all hell broke loose. Screams of 'Allah the Great' were countered with masterful outback obscenities. This disrupted von Kressenstein's plan as he had hoped to get further into the Anzac position and surprise them in daylight from behind. Too bad. The alert sentries foiled him and the battle started in the dark six hours before he really wanted. This ensured his troops were caught in the heat of the next day and exhausted them quicker. Of course, the Anzacs would also have to contend the heat and thirst:

> Our infantry were digging in whilst we held the line, then when our guns came on the scene, we retired, gradually falling to rear of the entrenched infantry, thus causing the enemy to advance & fall into the trap already laid for them.[76]

---

[74] Gullett, p. 190.
[75] MacMunn and Falls, p. 184.
[76] Birbeck, Tpr G., diary.

For two and a half days the battle raged backwards and forwards. Chauvel tried desperately to have his horsemen released to join him. Murray and Lawrence were obstinate. But Lawrence had little idea what was happening early in the battle. His telephone lines had been cut and he was so far in the rear he could not communicate with Chauvel or his own infantry commanders.

The British infantry commanders sat, refusing to join Chauvel without specific orders from Lawrence even though they had no other part in the battle. Brigadier E.S. Girdwood commanding 156th Brigade declined to involve his fresh infantry so that 1st and 2nd ALH Brigades could water and rest before resuming the battle. A couple of days into the battle and Lawrence established some contact with his brigade commanders and gave direction to his infantry on 5 August. Gullett records:

> On the first sign of dawn Chauvel had moved vigorously with his horsemen ... the infantry was then to make a strong advance ... and while the infantry pressed the right the cavalry was to envelop and crush his right ... Had the 52nd Infantry Division moved as the Light Horse had, this must have resulted in the capture of most of the enemy force ... but as on the previous day, there was no sound cooperation between the two divisions (Infantry and Light Horse). The infantry did not clear their defensive posts until 9am and had made practically no advance before 2pm. This delay was fatal to the whole project.[77]

Inexperienced, incompetent and snobbish British infantry leadership extended the fight and the war. In fairness, though, the poor bloody Tommy soldiers performed as best they could when they could. The 52nd Infantry Division had been in Sinai for a while and had some acclimatisation and sand experience but it was still hard going through sand by foot. The fair-skinned conscripts of the 42nd Infantry Division had no such experience. They carried massive weights in their equipment and hadn't

---

[77] Gullett, p. 164.

Battle of Romani

been taught water discipline in heat and succumbed quickly. At one stage, 800 men of the 42nd 'disappeared'; they just dropped out of the line, exhausted through lack of water and heat. It took two days to round them up. Sadly, this reflected on the British staff that they had not given sufficient acclimatisation or adequate water to unprepared troops – their own.

At near desperate times, two wonderful things did happen. First, the British Commander of 156th Infantry Battalion (not the 156th Brigade commander above) using his own initiative brought his troops into action to assist the exhausted horsemen who needed water for their horses and rest for themselves. Second, Lieutenant Colonel R.M. Yorke led a squadron of Gloucester Hussars, also on his own initiative, into action on the right of the Anzacs, thus deflecting a Turkish outflanking movement that could have destroyed the Anzacs. These two junior British commanders showed what could be achieved with cooperation, personal initiative and common sense.

Throughout this battle, the British artillery gave marvellous help to the horsemen, firing incessantly and accurately when called upon. Guns of the various territorial units and those of the Bing Boys met all the challenges that were asked of them. Similarly, even though their aircraft were inferior to those of the Germans, the airmen of the AFC and RFC flew continuously, returning to reload then out again; they directed artillery and naval fire and gave reports of enemy movements – invaluable stuff.

The stretcher-bearers were peppered as they recovered the dead and wounded. The medics went sleepless treating or evacuating. The vets were sleepless. So too the cooks and resupply drivers.

Among all this was a tight family of four. Unnamed in the reference, the father and his three sons were fighting as a Section. Dad was holding all four horses while the boys were engaged in shooting when a Turkish artillery shell landed in their midst. Three of the horses were killed, but dad was unhurt

Desert Anzacs

and later all four celebrated their good fortune.[78]

Lawrence finally released the Kiwis. Chaytor's Brigade arrived just in the nick of time. Murray then released 3rd Brigade. Chauvel was now able to commit his reserve regiments. The Turks were, as he'd expected, exhausted from heat and dehydration, as were many of his own. Then, with near half his force as casualties or prisoners, von Kressenstein withdrew the remainder. But the Anzac mounteds now were also exhausted, very thirsty (man, horse and camel) and hungry. Men were falling off horses asleep. Weak from dehydration and lack of food for three days, they had to rest before commencing the chase. The Kiwis and 3rd Brigade had to catch up and assemble. A chase was to be had, but not quickly.

Meanwhile Corporal Curran DCM, still under arrest from his Marseilles adventure, knew the stoush was on; he broke out. Being a man under arrest, he had no weapon. Nevertheless, he went forward to seek someone who would use his services; he became a stretcher-bearer at the field ambulance. Colonel Langley continues:

> He escaped from his guard, hating to miss a scrap and started out on his own as a stretcher-bearer. He found his way to the front line where he gave a drink and a cheery word to the slightly wounded men. The badly wounded ones he helped or carried in himself. Fourteen times he braved the flying bullets to bring in wounded men. His luck, however, petered out and on his next errand of mercy he was killed.[79]

As he had been under arrest at the time, no medal was awarded to Corporal Curran. William Massey, the British war correspondent at the time, reported:

> If Curran's relatives had no medal or posthumous award to remind them of a good soldier's work in the heat of battle, they may be satisfied his memory is a rich treasure and an inspiration to the brethren with whom he fought

---

[78] Gullett, p. 183.

[79] Langley and Langley, p. 47.

Battle of Romani

and for whom he made the supreme sacrifice.[80]

However, it was a changing world. For the first time, air warfare was being recognised for how its potential could develop. Pilots and observers of No. 1 Squadron AFC and No. 14 Squadron RFC maintained flights over the battlefield with the intensity of zealots. They developed tactics and skills even as they dodged the faster, better armed German aircraft doing the same thing; some didn't make it and went down. Their effort helped Chauvel determine his own troop movements and counteractions as well as prepare for the final attacks.

The integration of air support with ground troops was beginning to have a positive impact on ground commander tactics and results. The role of aircraft was becoming well appreciated and further understood by the more imaginative of both sides. And the airmen were learning how they could be of use to the ground. warriors.

But the Germans had both aircraft and anti-aircraft guns. The Allies didn't have anti-aircraft guns. Enemy aircraft, superior to the British aircraft, fought them off the battlefield at times then did their own bombing and strafing with minimal interference.

Horses and riders on both sides were devastated by this new menace of bombing and strafing, never before experienced. War was becoming even more unpleasant. On the ground, though, von Kressenstein had little option but to do as Chauvel had predicted. He could not defeat entrenched British infantry. He had to take the circuitous route to the non-entrenched Anzac mounted positions on the sand. He knew his Turkish troops were hardy and accustomed to the heat with minimal water. Never short on enthusiasm, Turkish soldiers would endure until the situation was hopeless, then surrendered willingly. Von Kressenstein had expected the advantage of surprise that a night insertion would provide. But he was caught out.

The Turks did have an occasional success too. At one stage the battle was so ferocious and uncertain that some of the light

---
[80] Massey, W.T., *The Desert Campaigns*, p. 74.

horse had retreated to the railhead. Here the cooks fed them while shot and shell fell all around where never a shot should be seen by cooks. Fed, watered and somewhat rested, the troops advanced and finally drove back the Turks.

Von Kressenstein used his mounted force and foot soldiers in a tactically sound and methodical way, nearly overcoming the depleted Anzacs at various points. But, just as Chauvel had predicted, the endless slow motion trudging through the sand, the ferocious heat while trying to fight a moving battle, the lack of water, the two timely interventions at the initiative of junior British officers, the stamina of the waler horse and Anzac leadership were von Kressenstein's downfall, despite Murray and Lawrence's inertia.

To give some idea of the heat, Trooper Bygott colourfully says, 'the sun would melt the sins out of Satan'. Water and thirst were issues for everyone.

In one dispiriting moment, some Johnny Turks had just been pulverised by crazy horsemen so up went white flags. Next thing, German machine-gunners turned their guns on the surrendering Turks. Seems Germans had a greater allegiance to their cause. Their significantly better healthcare, equipment, food, watering and leadership were likely to have contributed to that zeal. Racial disputes were a constant issue for the Germans and Turks.

At times the Turk showed moments of chivalry and decency towards his enemy. During the Romani battle Major Alan de Rutzen, Commander of No. 6 Camel Company, was shot while observing the enemy from atop a high sand dune. His body rolled down the sand slope and his soldiers were unable to recover it, the slope being in full view of the Turkish riflemen. His soldiers figured they could search after dark and recover him. This was attempted but failed: there was no sign of the body. Next morning however, a Turkish officer approached carrying a white flag. The first thought was the Turks would surrender; but no. The Turk was returning Major de Rutzen's signet ring, watch and some personal letters that had been

Battle of Romani

retrieved from a looting Bedouin as they thought his family might like them back. As an afterthought, the Turk mentioned they had shot the Bedouin.

In human terms, around 3,900 Turkish and German POWs were taken. These were the fortunate ones, as there was something in excess of 1,250 who would never again see their homeland and families.

Chauvel's management of his depleted AMD was outstanding. And he had good company. Universal Anzac leadership soon came to the fore. Lieutenant Colonel Scott (the major from the Jifjafa operation) and the 7th ALH Regiment were held up by a dug-in Turkish force. He had the choice to dismount his troopers and advance over the heated sand in normal light horse manner or remain mounted and charge the line. Choosing the latter, and pretending their bayonets were swords, his men suffered few casualties and won the engagement 'although under heavy fire – during the whole war the Turks shot badly if resolutely galloped at'.[81]

Then, Sergeant R.C. Sharp of the 9th ALH Regiment and his troop of around 30 horsemen swooped around the side of a much larger Turkish force and unexpectedly charged them with bayonet and rifle. Some 425 Turks surrendered with their seven machine guns to this outrageously small force.[82] Sgt Sharp later became Lieutenant Sharp and was awarded a Military Cross. This type of boldness by the Anzacs often won the day.

There is a quote by Ralph Waldo Emerson: 'What you do speaks so loudly I can't hear what you say'. No one saw generals Murray or Lawrence as they were so far behind the action. But they did see Lieutenant Colonel 'Galloping Jack' Royston, then a Regimental Commander with Chaytor's NZ Brigade. Hill reports his inspirational leadership:

> Galloping Jack was probably the best known and best loved in Anzac Mounted, a South African ... A huge man, he galloped from regiment to regiment, encouraging

---

[81] Gullett, p. 168.
[82] Gullett, p. 170.

the men by his cheerfulness and complete disregard of danger. It was said that he used up to fourteen horses on 4 August and that in action he looked ten years younger (he was almost 60). About 3pm Chauvel phoned Royston's HQ to ask how Chaytor's attack was going and was told: 'Colonel Royston is wounded and has gone for another horse'. Thereupon Chauvel rode off in chase and personally ordered Royston to a field ambulance to have his wound dressed.[83]

Meanwhile, the prosecuted Corporal Curran, until his death, and all the stretcher-bearers, had worked ceaselessly, day and night without sleep, to retrieve the wounded and evacuate the dead. Among the stretcher-bearers was Albert 'Tibby' Cotter, Australia's cricket test fast bowler whose cricket career was interrupted by this blasted war. The stretcher-bearers and ambulance drivers brought the wounded to casualty clearing stations (CCS) where the medics gave immediate treatment or arranged them for evacuation to be treated at the railhead and hospital. Neither fighters, carers nor repairers slept.

The Anzacs made two changes to the British system of casualty clearance. Firstly, stretcher-bearers were mounted to keep pace with the fighting troops whose wounded would need quick treatment – seems sensible enough. The British lads had to walk through soft, hot sand and somehow keep pace or catch up, then walk or carry their wounded back through the sand to the wheeled or animal ambulances. Not so sensible. Secondly, the British system of wheeled ambulances with a six-horse team (or in really hard going, up to ten horses) required three riders. The Anzacs put one driver on a buckboard, releasing two soldiers for other medical duties and treatments. Seems sensible enough. But guess what? The British wouldn't supply the different rein system because it wasn't their system; so the Anzacs had to put hand into pocket again and buy them from their soldier funds. Eventually, the British discovered this was a smarter method, even for their own wounded and provided the reins for all.

---

[83] Hill, p. 79.

Battle of Romani

Now, almost all of us at some time have had broken bones, major surgery, cuts and injuries in modern, sanitised hospitals that offer pain relief; still not a lot of joy. Imagine lying on a stretcher riddled with gunshot or shrapnel, with just a hat for shade in temperatures between 40^0C and 50^0C, with little food or water, minimal onsite treatment as the medics are so few in number and overstretched, with limited pain relief. But fortunately there's an empty train adjacent you pointed towards Egypt and the hospital. Then a British staff officer says the train is for the movement of Turkish POWs and the wounded will have to remain in the sun until the next train arrives. After all, they weren't designated hospital trains. Despite the howls of protest from Australian and British medical officers on behalf of the wounded, those staff officers had their way. It's possible someone might have wanted to shoot one or more of those staff officers. Despite knowing that a major battle was looming, the staff forgot or didn't think it necessary to order hospital trains. Many eyes closed for good, never to share another bushland daybreak or home beach sunrise with a loved one.

This dereliction of man-management and leadership by Murray, Lawrence and HQ staff fuelled Australian contempt for the EEF staff even further. Australian officers pleaded with our government to have our troops come under the command of Australian officers. That plea went unanswered.

The horses and camels weren't spared either. German aircraft, Turkish artillery and rifle or machine gun fire combined to test the Mobile Vet Sections. They too worked endlessly to treat or evacuate these animals that really were the backbone of the Anzac victory at Romani. The remount units in Egypt were feverish to keep up the reinforcements of horses. Clearly, it takes a great team to win at sport or win a battle. It takes top leadership, training, perseverance, discipline, common sense and pride when things get tough.

More than three days after the Turkish attack commenced, Lawrence finally gave command of all the light horse, New Zealand Mounted Rifles, cameleers, British cavalry and their attached artillery and machine guns to Chauvel. Lawrence then gave Chauvel the order to pursue. Instead of moments to assemble what should have been a concentrated force, it took hours to prepare the scattered troops; to feed and water men and horses, resupply the force with ammunition, water and stores then assemble the artillery, field ambulances, veterinary sections and signallers before a pursuit could begin. A concentrated force in the beginning would likely have seen the entrapment and end of the whole Turkish force on the spot and probably allowed the boys to go home earlier by a year or more.

Now, though, the chase was not an immediate prospect and many Turks were able to evade and withdraw eastwards to previously well-established and prepared defensive positions. They received food, water and medical care, were in fortified positions rather than unprotected desert, had access to letters from loved ones and generally felt safer. They recovered quickly.

The Anzacs and their horses were exhausted. Men slept in their saddles until they fell off, then remounted. Again, horses went without water for up to 60 hours. Men endured thirst with just one or two water bottles that lasted 24 to 36 hours in some cases.

Not everything worked well for the Anzacs. Brigadier Antill was leading the 3rd ALH Brigade, fresh from its release by Murray. With no previous battle or desert experience, having been held in reserve, this was the maiden battle for Antill and the 3rd. After their initial, brilliant execution of enveloping a Turkish formation and their collection of prisoners, the men were excited at their achievements and, although short of water, were quite fresh and keen to press on. However, these reformed regiments came under light Turkish artillery. Antill had them fall back rather than advance at speed. An advance by fresh troops could have completely routed the Turks. Chauvel was disappointed with Antill but he did recognise that same fighting spirit in the newly arrived force and was

Battle of Romani

encouraged for the future. By coincidence, a few days later General Birdwood in England requested Antill to command an infantry brigade in France. Chauvel made no protest, replacing him with Colonel Royston, whom he admired and promoted to Brigadier.

Once assembled, Chauvel's unrested force began a pursuit in earnest. The depleted and tired Anzacs marched on for six days, engaging in various skirmishes, exhausting the troops further. Sleep, water and food were taken when they could; rarely in comfortable circumstances, as Trooper Bygott relates:

> We watered our horses and got water for ourselves, but though it was terribly salty it was very refreshing. We now drew rations for the next day, which had been brought over the desert on camels. We would then boil our black billies and with a bit of bully, an onion and some biscuits, we would have our night meal. This over, you would get your saddle-blanket and lay it down on the sand which is now nice and cool and with your saddle at your head and a feed bag for your pillow, and while you are having a few draws at your pipe, who is true pal to you, you wonder what the morrow will bring forth.[84]

The Turks, however, had received reinforcements and showed their resilience and clever use of their reserve troops. Once more the Turks demonstrated how well they could fall back to prepared positions and fight another day. They counterattacked:

> The Turk was ready and full of fight; he was far more numerous and more strongly supported by artillery. When he had measured the strength of the attack and brought it to a halt, the Turkish commander counterattacked; Chauvel's troops had done all they could and were forced onto the defensive ... he ordered a general withdrawal ... this was the signal for the Turkish infantry to move in for the kill. Only fine leadership and steady discipline saved the weakened regiments; it was a near-run thing, with the Ayrshire Battery nearly losing

---
[84] Bygott diary.

Desert Anzacs

their guns and the NZ machine-gunners pouring fire into the enemy at 100 yards.[85]

The counterattack put 6,000 Turks against Chauvel's force of 3,000 and the Turks were well rested. Their attack nearly came off, except for the well-disciplined extraction and excellent leadership again displayed by officers and NCOs.

Chauvel engaged the enemy with strong patrols at their flanks, knowing that their supply lines were severely stretched and they could not hold on for long; even with their numbers their position was not tenable over the long term.

Sure enough, within days the Turks withdrew into El Arish, 50 miles away, at the edge of Palestine. The pursuit was over.

Romani was won. The Anzacs had brought about a stunning victory.

But there'd been a price. Most mere mortals feel pretty shattered after a late night out or a long day at the office; worse if it's a night without sleep, or jetlag after a London to Sydney flight. But by the end of this battle most of the mounteds had been in the saddle up to six days with little sleep while moving, riding, fighting, running through sand in the furnace, unfed, barely watered and with many of the horses dehydrated. They were shattered. But amazingly, a man's willpower can be unfathomable in times of crisis – and makes the difference between survive or perish. Australia's *Official History* reports:

> Gaunt from prolonged sleeplessness, their eyes bloodshot from glare and strain, their faces begrimed with dust and sweat, and bristly with a few days growth of beard, the Australians and the Wellingtons might have unnerved troops in better condition than the unfortunate Turks opposed to them.[86]

And the horses:

> These wonderful Walers were so exhausted ... despite all their spirit ... they lay down in the sand at each temporary halt, but when urged by their riders, responded gamely

---

[85] Hill, p. 81.
[86] Gullett, p. 165.

and carried them forward ... Their capacity to suffer and continue working was unsuspected even by their Australian riders.[87]

Von Kressenstein was commended. Although failing its objective, the combined Turko-German force made an amazing advance, attack and withdrawal then a counterattack, all performed with great staff work, leadership and endurance. They just had no answer for those crazed Australian and New Zealand horsemen.

Romani was the first major British victory in two years of the Great War. Delivered by Chauvel and the Anzacs. It was the turning point of the British campaign in the East. The Suez Canal was never again threatened.

The Anzacs stood and had been counted; men and horses, planes and pilots, riders and camels, vets and medics, signallers and cooks. Accolades flowed from Australia's political and military leaders. But how could this be; a bunch of bushies with no serious wartime experience compared with centuries of British military imperialism and traditions? Murray too gave praise.

However, this was offered to the War Office and public as a British victory. Few in London really understood Murray's lack of ability due to his self-protective battle reports to cover his incompetences.

First, who was the battle commander; Murray or Lawrence? Both had reserve forces but both were in distant locations. Both were out of touch with commanders in battle. Keogh remarks:

> Although the EEF won the battle of Romani, the engagement reflects little credit on the commander responsible for its conduct.[88]

He goes on to say the distribution of key infantry in an area where no threat was foreseen was 'hard to understand'. In effect, they could not and did not contribute to the battle.

---

[87] Gullett, p. 174.

[88] Keogh, p. 56.

Desert Anzacs

Worse, they refused to join Chauvel's effort when doing so could have brought it to a quick end.

Murray's separation of the brigades and Lawrence's retention of mounted troops seems almost childish in its conduct. Fortuitously, General Chauvel knew what he was doing, as did his brigade commanders and soldiers. As Gullett reports:

> The High Command did not excel at Romani, but the result was still a splendid and far reaching triumph for British arms. And, considered from any angle, this triumph must stand almost entirely to the credit of Chauvel and his Anzacs.[89]

Of course, it was difficult for either Murray or Lawrence to appreciate what was happening once battle commenced. They were both so far in the rear and communications failed. Keogh continues:

> There was no overall coordinating commander in the battle zone. General Lawrence's plan to control the battle from Kantara was faulty to an extreme degree. When the telephone line was cut he might as well have been in Cape Town (South Africa) as in Kantara. From that moment he lost control of his battle and exercised no influence whatever over the course of events until the crisis was past.[90]

Lawrence had repeated his failures of command and control that he and Wiggins had committed a few months earlier when the 5th Yeomanry was wiped out. Although Murray could see the big picture he had no idea how to make it happen.

Murray's next effort surprised – no – astounded everyone. He had verbally heaped praise and appreciation upon the Anzacs for their victory. But his official dispatch to the War Office, which was subsequently published in global newspapers, credited the British infantry with events that they had not been party to. He ignored the Anzac mounteds. The casualty list gave clear indication as to who had been really

---
[89] Gullett, p. 190.
[90] Keogh, p. 57.

Battle of Romani

involved. No one could have been more astounded than the British infantry who must have thought Murray had watched a different movie. Certainly, the Anzacs knew he had.

Gullett records:

> Still more difficult to understand was the discrepancy between Murray's messages of appreciation to the troops and his list of awards to officers and men for service covering the period of the Romani fighting. The great majority of these went to troops recruited in the United Kingdom, and an excessive number to the officers of the Staff that had blundered in the conduct of the fight from beginning to end. Had no awards been made, the Anzacs would not have complained; but the publication of a list so discriminating and unfair caused much discontent.[91]

Soldiers don't fight to win medals. But when an undeserving group is awarded in preference to those who do deserve it, discontent and distrust abound. It's like the drug cheat who takes the Olympic gold at the expense of the honest athlete who takes silver, or fourth. British favouritism of their own plagued the Anzacs throughout the Sinai Palestine campaign as it did in France – it bred contempt and resentment during and after the war.

Others knew that Chauvel and the horsemen created the victory. Britain's Colonel Wavell, Murray's Chief-of-Staff, described the Anzac horsemen this way:

> Endowed with a natural aptitude for the work and a fine physique, the Dominion horsemen soon became seasoned warriors, and from now til the end of the war did magnificent work. As now in the desert, so later in the steep, rocky hills of Judea and Moab, they showed the value of enterprising horsemanship even in terrain the most unpromising for mounted troops.[92]

To quote Chauvel:

---

[91] Gullett, p. 192.

[92] Wavell, *The Palestine Campaign*, p. 45.

> It was the empty Turkish water-bottle that won the battle ... Romani was the first decisive victory attained by British Land Forces (excluding the campaigns in West Africa) and changed the whole face of that campaign in that theatre, wresting as it did from the enemy the initiative which he never again obtained. It also made the clearing of his troops from Egypt a feasible proposition.[93]

Sometimes, little things have big meaning. In his biography of Chauvel, A. J. Hill relates what could be thought a minor incident. While riding, Chauvel and some of his HQ staff came upon a solitary trooper whose horse was down and exhausted. The weary trooper had poured water from his own water bottle into his hat, hoping to help the horse drink. Chauvel dismounted, added his own water to the hat, then looked expectantly at his officers; they took the hint and did likewise. This trooper had a big tale to tell his mates that night. Minor things can give a massive boost to morale.

In addition, in a soldier-to-soldier moment after the Romani battle, Chauvel visited each of his regiments to express his admiration and his gratitude for their remarkable efforts. He had a gift for understanding his men, which is the essence of leadership. Hill also reports the respect his men had for him, made plain in a letter from a young trooper to his father after Chauvel had returned from leave in England:

> Yesterday we were inspected by General Chauvel. We are all pleased to see him as he is well liked by both officers and men. A chap feels pretty safe with a leader like him. I saw him riding backwards and forwards under heavy fire at Romani and it seems he did not know what danger was.[94]

Romani was won in August and the Turks withdrew into El Arish. September, October and November 1916 saw little fighting. It was autumn so there was some relief from the intense heat. It was a period of recuperation, resupply and

---

[93] Hill, p. 83.

[94] Hill, p. 86.

preparation for an advance towards Palestine.

In these three months, the Anzacs continued their mounted patrols while the airmen gained invaluable experience through recon and photo runs, air-to-ground communications, the odd bombing mission, and innovations to keep their flying machines flying. Skirmishes with Turkish mounted patrols and isolated detachments kept the boys on their toes. The troops continued to master navigating in both daylight and under the canopy of stars. The British Yeomanry improved their desert skills too, giving them much-needed confidence. Yet, life between battles wasn't a picnic. The strain of not knowing what was around the corner was ever present, as Robert Bygott recalls after one skirmish:

> It is dark now and you can still hear strings of gurgling camels going past, the wounded are being brought in on cacolets, sort of stretchers on either side of a camel, you realise that they have to travel miles through desert in this miserable way before they reach a decent hospital and their pain and suffering is hard but still one looks upon them as lucky and a certain sort of envy as some have finished with war now and will go back to Australia and others who are not hit so badly will be out of trouble for a month or so, but you yourself have still to go through it, and you can't look ahead. Then you have the other poor chaps who are by this time under the cold sand, their worries and troubles are over, but one misses them terribly. One realises what a game of chance this is, some are lucky and some unlucky.[95]

But there was some time for rest away from the action. Letters to loved ones. Diaries to be entered. Sores and injuries to be treated. Energy levels to be topped up. Human needs of tenderness and romance were missing and thoughts of home ruminated. Letters from home were eagerly awaited. After all, most of these troopers were still teenagers, with less than mature natures and desires. Even the older ones were human – something we overlook when discussions of soldiers take

---

[95] Bygott diary.

place. Even enemy soldiers were human with similar thoughts and feelings, compassion and emotions. In the desert, unlike the fields and modernisation of Europe, soldiers of both sides had nothing to break the monotony and boredom of the desert, the heat, the bugs, the boring food or the spartan skyline. The importance of letters from home, a cheery note, a family photo, expressions of love and kindness were more valuable than gold.

By October 1916, Murray decided his varied responsibilities as overall Commander of the Egyptian theatre, including how to deal with the peripheral Arab Revolt, required him to move his HQ from the canal to Cairo and a suitable hotel. He appointed Major General Sir Charles Dobell to command what was renamed the Eastern Force, being all the forces on the canal plus those in Sinai, and promoted him to lieutenant general.

He also asked for an officer of lieutenant general rank to assist Dobell and to command the troops in Sinai. Philip Chetwode, a cavalry officer, was sent from France and the Desert Column was formed with the Anzac Mounted Division, the British Yeomanry and the camel brigade.

It's easy to imagine there would have been great joy amongst officers and men when Chetwode replaced the disappointing Lawrence. Transferred to France and posted to a headquarters somewhere far away from soldiers and any tactical requirements, he reportedly did a splendid job in a desk role. Too often, too late, does war (or life) find leaders an alternative hole from the positions in which they should never have been placed. Things improved hereafter, rapidly.

By now there were around 150,000 British, Australian and New Zealand troops and 6,000 Indians in Sinai. No one knew for sure how many Turks. The road to El Arish was opening. Murray knew that to keep the Turks away from Egypt and the canal he had to move from the Sinai into southern Palestine and at least hold that line. Thus, El Arish became the first objective. The War Office now agreed.

In this setting, the railway, pipeline and telegraph

Battle of Romani

progressed, albeit slowly. As did more than a couple of hundred miles of wire meshing laid as a roadway in two directions to ease the passage of foot soldiers, artillery and wheeled transport for baggage and stores. The pipeline was providing one and a half million gallons of water a day, most of which was needed by the tens of thousands of labourers on these projects. However, the horsemen and cameleers scrounged whatever they could to supplement their measly one or two water bottles per day.

This was also a period when the soldiers could take some well-earned leave and head to Egypt and Cairo. Lance Corporal Burgess wrote the following home:

> This afternoon Joe Crouch and I got permission to ride over to the Citadel. It took us 1¼ hours to go. The Citadel looks a strong place with great walls all round it. Right in the centre on top of a hill is the Mosque, it is said to be the best in Egypt and second only to the one in Constantinople. It is called the Mosque of Alabaster for inside the walls are entirely composed of alabaster, all except the four great pillars which are only imitations as alabaster is not strong enough to carry the weight. The ceiling is beautifully decorated. The guide told us the man who did the work had his eyes put out when it was completed so he could not do any work like it again.[96]

It must have given young Burgess and his mate Joe quite a cultural shock to learn how good work was rewarded.

By the end of November 1916, General Murray considered that the Turks had three choices: to go home; to attack and pursue their objectives to capture the Suez Canal and Egypt; or to stay and defend where they were. No one could say for sure what they would do. The War Office and Murray reached an agreement. El Arish should be the objective and then a move into Palestine considered. But no extra troops could be spared for Murray's endeavours. He would have to make do with what he had.

---

[96] Burgess, letter to family.

Dobell and Chetwode were ready to move by the end of November.

Now the Turks were making mistakes just as Murray and Lawrence had done previously. They had smallish groups defending isolated positions throughout northern Sinai. Extensive trench systems and fortifications gave them some confidence and they had what they thought were adequate artillery and machineguns.

There was only desert and these few Turkish outposts between the Desert Column at Romani/Katia and their objective, El Arish. There was no water. Aerial reports and local intelligence had indicated there was a force of around 2,000 well-entrenched Turks in El Arish where there was water. Twenty-five miles southeast was Magdhaba, which defended the end of the Turkish railway from Beersheba and supported their troops in the Sinai. Both had to be taken. And after that, the door to Palestine would be ajar.

For the Desert Column to advance, enormous quantities of water would need to be carried by camel as the pipeline had not advanced that far. And the further they moved away from base camp, the further they moved from their food supplies. Keeping the infantry and mounted troops fed adequately was a logistical nightmare and many went desperately short. It took until 20 December to accumulate the water and stores necessary to approach El Arish. Finally, a night march was planned for an attack the next morning.

Aerial reports on the day indicated the Turks had abandoned El Arish. The Desert Column simply occupied it without a shot; one battle avoided. The little village, approached over ground that became progressively firmer after the sands, had palms for shade and water. But, like many Arab villages, it was filthy and squalid, without basic cleanliness and hygiene. Its stench was totally repugnant to western noses and stomachs. It was also infested with sickness and diseases to man and horse. The medical staff had to chlorinate the water and inspect the hovels for sanitation. Meanwhile, the Arab inhabitants let out their

Battle of Romani

unique, high-pitched squeals of delight as the Anzacs came into their earthen laneways, throwing flowers and pulling at trouser legs as was the Arab custom of welcoming horsemen. It's unlikely that the men appreciated their horrid smells and dirt-encrusted hands as much as they did leaving the sand and finding firm ground. Brigadier Charles Cox, commanding the 1st ALH Brigade, wrote:

> That night will always seem to me the most wonderful of the whole campaign. The hard going for the horses seemed almost miraculous after the months of soft sand; and, as the shoes of the horses struck fire on the stones in the bed of the wady, the men laughed with delight. Sinai was behind them.[97]

After that night march, their eyes fell upon patches of green; young barley, fruit trees and other plants and trees. Averting eyes from the despicable village, imagine this picturesque sight after nine months without water, of brown landscapes, of sucking sand, of bugs and critters.

El Arish was theirs but the job wasn't done.

The Turks had retreated to Magdhaba to the southeast and Rafa further north. These were the last two Turkish outposts in Sinai and the commander wanted to take them out to protect the flank. With that, General Chetwode tasked the Anzac Mounted Division to attack Magdhaba.

Chauvel had all but one ALH brigade. He was given command of the various camel companies that had been formed into a brigade that remained under his command for the rest of the campaign.

Now though, the detail of the fight itself is not the issue. Some of the related stories are of interest to show what happens in battle.

Firstly, during the night march, an infantry brigade of the Scots and Chauvel's baggage camels met and their paths tangled. The traffic jam took some time to untangle and delayed

---

[97] Hill, p. 87.

Desert Anzacs

progress in the dark. Luckily, they didn't shoot one another by mistake.

Daylight and No. 1 Squadron AFC, now with Major Dickie Williams as squadron commander, was in the air. At 6.30am they were bombing and observing the village and Turkish defences. Up there, as a fascinated, unofficial observer, was Lieutenant Colonel Guy Powles, a New Zealander and part of Chauvel's HQ. His observations were quickly converted to a note and dropped to Chauvel; it read: 'The bastards are there alright!'

A welcome feature of the presence of No. 1 Squadron and the firm ground was that the airmen could land close to Chauvel's HQ and bring immediate verbal reports and sketches of enemy actions and dispositions. This instant information gave the ground troops a greater appreciation of the airmen and their capabilities that has since developed into the modern use of satellites and drones.

However, this time the Turks had made a tactical mistake by leaving an isolated outpost in the open, even though it was well entrenched. Yet, the real enemy was water. The day dragged on and Chauvel knew he would have to withdraw to water his men and horses. He ordered the withdrawal late in the afternoon, unaware that Brigadier Cox's 1st Brigade was making progress. When that order arrived Cox told his signaller, 'Take that damned thing away and let me see it for the first time in half an hour'.[98] A few minutes later 'Fighting Charlie's' regiments had captured most of the place. Just then, too, Major Robertson, 22 years old and commanding the 10th Regiment, had his troopers surround a much larger Turkish position and forced them to surrender. Magdhaba was taken, once more with boldness and dash, just in the nick of time before water forced a retirement. And Brigadier Royston just couldn't help himself:

> Royston, who, as usual, was riding about in the thick of the fight, attended only by his orderly, galloped up

[98] Hill, p. 89.

Battle of Romani

to a Turkish trench and was instantly covered by five Turkish rifles. The old fighter excitedly raised his cane and, knowing no Turkish, shouted at the riflemen in Zulu; whereupon the Turks, impressed with the demonstration, dropped their rifles and held up their hands.[99]

It was Christmas Eve, 1916. The main force returned to El Arish, leaving troopers to protect the field ambulance that was still collecting and treating the wounded. The airmen returned to their airfields to await the next sorties. The rest of the AMD rode into the night where Hill reported strange goings on:

> Few were the men or horses who had slept in the last three nights; now men slept in the saddle in spite of the piercing cold, as dust rose in stifling clouds and stumbling horses collided in the dark or fell with their riders. It was misery mitigated by snatches of sleep. For some, their sleep was punctuated by strange hallucinations; men told later of visions of streets and well lit houses and strange animals. Those who saw it were amazed when Chauvel and another officer suddenly galloped off to a flank; after a mile or more they reappeared and took their places quietly in the column. Later each confessed he had imagined himself to be fox hunting and had galloped after the fox.[100]

On Christmas Day, 1916, life, water, food and sleep were the soldiers' gifts. General Chetwode was a true soldier. He understood what the AMD had been through and achieved. Hill reports:

> Chetwode signalled to Dobell: 'Conduct of action was left entirely to General Chauvel. I consider that he conducted it most admirably under most difficult circumstances and if you concur hope you will bring him to favourable notice of C-in-C'.[101]

---

[99] Gullett, p. 224.
[100] Hill, p. 89.
[101] Hill, p.89.

Desert Anzacs

The last outpost before Palestine was Rafa. Another Turkish error left 2,000 troops isolated there to be surrounded completely by the Desert Column. General Chetwode moved forward to direct as a good commander should, with greatly improved results.

There were still problems for the attackers. An open approach with no shelter, no water, heavily defended with entrenchments and cover for the defenders, and, this time, enemy reinforcements on the way. The detail of the fighting doesn't matter too much, other than Chetwode gave the order to withdraw as the position seemed insurmountable at day's end, water not being available. And the Anzacs were already engaging the reinforcements at a distance. Despite seeing the order to withdraw, Chaytor and his New Zealanders decided to give it one more shot. Lieutenant Colonel Langley was of the same mind as Chaytor, leading his camel brigade in for a final charge. With bayonets glistening in the last of the sunlight, both forces charged, screamed, fired and scared the daylights out of the Turks who threw up white flags all around their position. A startled Chetwode, heading home on his horse, made a hasty turn around to find his force gleefully occupying Rafa with nearly 2,000 prisoners.

Rafa itself was no major prize, other than clearing out the Turks, but it did provide some firsts. The 7th Light Car Patrol, a British unit under the command of Lieutenant W. McKenzie, a New Zealander as it happened, came into action for the first time. Sporting machine guns they were able to give effective support to the mounteds while they were themselves on the run. Light cars, including the 1st Australian Light Car Patrol under the command of Captain E.H. James, would later prove invaluable as fire support and reconnaissance. In this instance, at the end of the engagement, McKenzie had the guns removed and assisted in the evacuation of the wounded.

The field ambulances and two brigades of horsemen remained to recover the wounded and drive off looting Bedouins. Throughout the night they also provided medical

Battle of Romani

care to the wounded Turks. All treatment and evacuation worked smoothly this time, as adequate preparations had been made.

AFC and RFC airmen had remained overhead most of the day. For the first time in battle they were able to use wireless to direct artillery and ground troops. They were also able to take photos and land them quickly for invaluable map-making on the spot.

Of further interest was the comparison of casualties. You'd expect a force attacking up hill in open country to have more casualties than the protected defenders; but that wasn't the case. Although 71 allied troops were killed and 415 wounded, the Turks suffered 200 killed, 168 wounded and around 1,600 prisoners. This highlights that marksmanship was not a Turkish forte. It also highlights that, under ANZAC command, our troops weren't just told to charge and to hell with the consequences, quite unlike what the British commanders were doing in France. It further shows that boldness and dash can cause the recipients to act erratically and capitulate.

Outside of the military, inspired leaders and noble workers appreciate the value of recognition of a job well done. In some instances, a pat on the back with enthusiastic words or a handwritten note is enough. Sometimes complete glory is acceptable, like a Nobel Prize.

Chetwode understood military recognition. The day after the taking of Rafa he issued a Special Order:

> I am intensely proud of the splendid behaviour of the Mounted Troops and the ICC Desert Column in the severe and successful operation of yesterday at Rafa. The position was a particularly strong one with almost glacis-like slopes and defended in a most determined manner. I heartily congratulate and thank all taking part in this very successful affair.[102]

For soldiers, recognition comes in the form of honours and awards, often delivered as part of a New Year Honours or

---

[102] Hill, p. 94.

King/Queen Birthday List. Sometimes, a commander receives an honour that symbolises all the men under his command in recognition of their collective achievement. The British system is stratospheric in pageantry and awards, for their own.

Chauvel had been passed over by Murray for an award for his actions at Romani. This was a slight not just to Chauvel but to the entire Anzac force that brought about that victory. No award had appeared for Chauvel in the New Year honours list of 1917 but in mid-January, in a moment of magnanimous posturing, Murray wrote to Chauvel advising that

> His Majesty has graciously acceded to my recommendation that your services and the gallantry of your Division at El Arish, Magdhaba and Rafa should be recognised by appointing you a K.C.M.G. I send you and your Division my sincerest congratulations.

This was a significant knighthood to follow on from Chauvel's CMG from the Boer War and his CB from Gallipoli. But it opened a can of worms over Murray's favouritism of his ineffectual infantry at Romani and the neglect shown to the Anzacs. The collective opinion of Murray went from low to subterranean in quick time.

Mention must be made of Bedouin behaviour that reflected the lore of the desert and indigenous tribesmen with morals and culture that were different from the Western soldier:

> At dawn the Bedouins swarmed over the field; an hour later it had been stripped as clean of fighting material as the surrounding country. Uniforms and boots were torn savagely from the dead, even from the Turkish wounded who still remained on the ground; afterwards it was found that even the graves had been opened by these wretched natives in their lust of gain. All through the campaign, British policy pandered foolishly to these degenerate roaming Arabs of western Palestine. Firm punishment at the outset would have saved infinite trouble later on and the loss of many good Australian and British lives by murder. But the Foreign Office, entirely ignorant of the quality of these people, insisted that the Army leaders treat them as respectable

practicing Moslems; instructions were given that special care must be taken not to offend their susceptibilities. The Bedouin who were almost entirely without either religious or moral principles readily took advantage of the situation.[103]

In January 1917 it was time for another replacement of troops. The War Office recalled the now experienced 42nd British Infantry Division to France and replaced it with the 53rd Territorial Division fresh from Salonika under command of General Dallas. The 54th Infantry Division with little battle experience was brought from the canal. In addition, 4th ALH Brigade was formed from three regiments that had previously been in Gallipoli and had been under employed in Egypt. This excited Chauvel. He placed Brigadier Meredith in command of the 4th after his outstanding performance at Romani commanding 1st ALH Brigade.

General Murray soon dashed Chauvel's excitement. In a fit of pro-British and anti-dominion sentiment, he once more split the Anzacs. He formed two mixed mounted divisions. What he called the Imperial Mounted Division took two of Chauvel's brigades, the 3rd and 4th, and combined them with two British yeomanry brigades, the 5th and 6th, and placed them under the command of the Brit, Major General H.W. Hodgson, himself a good bloke and well-respected by the Australians. The term 'imperial' was recognised to mean 'British', a term that was vulgar to the predominantly numbered and vastly more capable Australians.

Despite the strong Australian presence, the divisional staff was British, with little to nil experience of battle. More importantly, they had little experience with the battle-hardened Australians among them. In justification, Murray wrote to General Robertson the Chief of Imperial General Staff (CIGS) in London (Britain's top soldier at the time) advising that 'There had been a time when the yeomanry always got lost and the

---

[103] Gullett, pp. 242-3.

Anzac cavalry had to find them and bring them home'.[104] But, he claimed, that was some months before and since then the yeomanry had become more desert-worthy and battle prepared: 'Now they are learning to find their own way about'.

That left the Anzac Mounted Division with the other two ALH Brigades, the 1st and 2nd, with the New Zealand Mounted Brigade and a British Mounted Brigade, who were, at least, under the command of General Chauvel. However, Chauvel was again denied Australian staff officers and handicapped with more inexperienced Brits.

Good old Murray insisted on, and repeated, his Romani debacle, by splitting his Australians resulting in two weakened teams rather than one great one and one OK. Chauvel was unprepared to challenge, as to keep the Australians together would have sacrificed the Kiwis to British command – he would not do that to Chaytor and the Kiwis so the reorganisation stood. Even worse for Chauvel, was the presence of Brits on his HQ, since they were even less capable than the yeomanry. He challenged this vociferously but Murray remained obstinate.

While the Anzacs developed a friendly regard for some British officers and soldiers, their contempt for Murray's favouritism over Anzac officers knew no bounds. Even the British officers were dumbfounded and realised they were out of their depth as operations developed. And Murray's poor tactical leadership made his presence more unwelcome:

> The evil results from such a policy were obvious and many. In Palestine, the army in early 1917 was staffed by officers of whom many were low in capacity. Highly capable officers in the Australian and New Zealand brigades were ignored. Hodgson's divisional HQ included nineteen officers, of whom, at the outset, two, Captain W.J. Urquhart, a junior staff officer, and Major F. Murray-Jones, the veterinary officer, were Australians; the rest were British.[105]

Murray had succeeded once more in infuriating the best

---

[104] Hill, p. 96.

[105] Gullett, p. 256.

Battle of Romani

troops at his disposal and confirming his total lack of leadership and comprehension of commanding troops of different nationality and capability. The protests of Chauvel and his brigade commanders to General Birdwood went ignored as he refused to be involved with local tactical decisions. And the Australian Government remained powerless.

Nevertheless, the tough ordeal of the Sinai campaign was now behind them . The desert, lack of water and food, wind and dust, flies and other creatures, were certainly not missed by the soldiers. The soft sand had been slow going and it had been almost impossible to use wheeled transport, hence the horses and camels in the thousands provided the transportation needed for fighting and stores.

The stamina of the horses was well beyond anyone's wildest dreams. The camels had proven vital and would be retained until the preparations for the final assault of this campaign for stores transport. Fighting was followed by periods of consolidation, not brought about by the enemy but by the weather and the immense logistic need to bring everything into the desert, especially water. The railway and pipeline slowly crept forward. The medical services were outstanding and their innovations lifesaving. The vets kept the animals fit and task-worthy. The airmen, starting with little idea of what to do, quickly adapted, developed and are to this day indispensable.

All this frustrated the enemy. German Head of Military Mission, Liman von Sanders reports:

> The British made their rate of progress ... entirely dependent on the establishment of secure and efficient communications (that included the rail, telegraph, pipeline, wire roads). It was the opposite of the method adopted by Turkey, which in all plans of operations underestimated the supply difficulties. To meet this slow but sure advance of the enemy Turkey should have brought up good fresh troops. But there were none to be had, for Turkey was tied up with Galicia, Romania, Macedonia, with two armies in the Caucasus, in Iraq and in Persia, and the Fifth Army had been made into an army of depot troops. With no training and their lamentable

condition the armies at Adana and Alexandretta could not be considered at all.[106]

Topping all that off, sickness and disease was a constant problem for the unhygienic Turks whose camps would often be found spread with decaying animal carcasses, rubbish, uncovered latrines and infestations of flies, fleas and lice, not to mention the nose-grabbing stench of rotting flesh and waste that seemed not to offend Turkish nostrils. Desertions were an endless issue. Nevertheless, in those circumstances, von Kressenstein had proved himself a worthy adversary in planning and execution; too bad he didn't have the political support he needed

For all the criticism of General Murray, it's acknowledged that his EEF had protected Egypt and driven the Turks from threatening the Suez Canal. His involvement in its execution did him no credit, but his plans worked thanks to others.

Sinai was now just bad memories; the sand, heat and waterless brown waste were all just horrid memories. The door to the Promised Land was now ajar and being pushed. The combined Anzac and British forces were riding and marching on firm ground, surrounded by green, wild flowers, veggies, orchards and watered pastures among the Arab villages. Anzac mouths tasted sweet fruit plucked from trees as they rode by and their horses chewed grass for the first time since leaving the Nile delta ten months earlier; happy horses make happy riders.

The Turks fell back to a line between Gaza on the coast and Beersheba about 25 miles inland.

As the railway and water pipeline crept onwards, the telegraph line progressed. Forward airfields were being constructed. The Royal Navy brought supplies in by sea, while the Royal Australian Navy's Bridging Unit built a pier and was unloading stores over the beach at El Arish.

Wheeled transport could be used for the carriage of supplies and as motor ambulances. Artillery teams could be reduced from eight or ten horses to six or four or, motor lorries

---

[106] Von Sanders, p. 145.

could pull their guns. Motor lorries reduced the numbers of baggage camels and attendant native drivers, which meant a reduction of demand on logistics and water. The light car patrol and armoured cars could make haste in reconnaissance. The mounted troops could now gallop rather than walk, better delivering their natural boldness of action. The British infantry, now acclimatised and with some battle experience, could make better progress and play a more effective role.

By January 1917, more than two years after war's start, France was still in stalemate with mounting death and destruction of the French countryside, villages and towns. British and Australian public morale was suffering with the casualties. Romani, El Arish, Magdhaba and Rafa had given some hope. Reinforcements were needed but volunteers were stepping back, not forward, having seen the destruction in France but being largely unaware of Sinai and its successes. There were no war correspondents, journalist or photographers; no visits by politicians or military leaders to tell this story. In Australia, conscription was discussed but rejected in two referenda. England conscripted.

British Prime Minister Lloyd George was butting heads with his military chiefs. He wanted to go for the Holy Land and achieve morale-boosting victories in the east. Others wanted to secure France and the Western Front. The politicians ruled and the military acquiesced. Murray would receive additional territorial infantry and artillery and increased Indian cavalry. But he would give up some of his now desert-experienced cavalry and infantry. He could enter Palestine.

From January to March 1917, on the doorstep of Palestine, preparation started for the advance of the EEF. The Australian airmen were tasked to observe and report enemy activity as well as create as much havoc as their slower and less capable planes could. No. 1 Squadron AFC and No. 14 Squadron RFC flew over daily. They dodged 'archie' (anti-aircraft fire) and the German Rumplers with their through-the-propeller machineguns. The threat was not only physical injury or death. This totally new warfare played psychological tricks with pilots' minds.

Lieutenant Mills was an observer in Lieutenant Roberts' plane on 24 January when they were hit by archie fire. Mills received a shell splinter through his wrist that exited at his elbow, severing an artery. A makeshift tourniquet and Roberts got him home for treatment and physical survival. The aircraft was riddled with holes and broken parts. Mills recovered fine. Although Roberts escaped injury, his mind was troubled. Sleeping difficulties struck him nightly; on a mission a week later and exhausted at 6,000 ft he fell asleep at the controls. Archie fire woke him in time. On his return, the medical officer removed Roberts from flying duties and he was repatriated as the first psychological casualty of the Flying Corps. Others would follow.

Little was understood, perhaps still isn't, about the workings of the human mind. The effect of warfare on soldiers from the Vietnam conflict are still not properly understood and what was called shell shock in 1918 is now called post-traumatic stress disorder (PTSD), but no one knew how to treat it back then.

Two incidents could have ended disastrously but fate stepped in to ensure that the RAAF would be born and would grow into a fine adult.

On 4 March, Dickie Williams was leading a flight of six mixed British and Australian aircraft. A piece of archie gashed a hole in his wing so he dropped his bomb load and turned for home. In short time, he realised his engine had stopped. Much pushing of starter buttons, quiet conversations with the engine and a check of petrol supply had no effect. A crash landing was prepared and as he got lower his alarm was heightened as bullets whizzed around. Dickie never got upset. He knew that the mail that only came each six weeks was due tomorrow and he didn't want to miss it. While preparing for his rough landing he recalled a previous incident with his mechanic – the ignition! It was easily turned off if the pilot didn't move carefully. With a glance at the rapidly approaching ground, he checked the ignition switch: off. Flick, then a belch of black smoke and the future 'father of the RAAF' and Air Chief Marshall, was once more heading skyward and homeward, a happier man. After that the mechanics fitted a metal guard over all ignition switches and Dickie got his mail next day.

Then, on 20 March, four aircraft were on a bombing mission. Lieutenant Frank McNamara (Macca) tossed a bomb out that immediately exploded, sending shrapnel though the undercarriage, his seat and into his buttocks and thigh. Firing smoke flares to tell his mates he'd been hit, he turned and aimed for home. On the way, he spotted Captain Doug Rutherford's plane on the ground and lots of Turks heading his way. Macca landed beside Rutherford who leapt on the wing as their mates in the two remaining aircraft strafed the advancing Turks, some 80 to 100 yards away. Macca fired his revolver and attempted take-off. But Rutherford's weight upset the balance and the aircraft broke up. The less wounded Rutherford carried Macca to his plane, plonked him in the pilot seat so he could press the ignition as Rutherford turned the propeller. It started first go. On board, and they were bumping through the sand. Somehow it got airborne, with blood oozing all over the place. An hour and a half later McNamara landed at home base where he immediately collapsed and was taken to hospital where things got worse. A tetanus shot given by the doctor produced an immediate allergic shock and he collapsed into a coma. Eventually resuscitated, he recovered. Two notable things happened. McNamara was awarded the Victoria Cross for his action; the only Australian AFC officer in the whole war to be awarded that honour. Second, after the war he went on to become Air Vice-Marshall of the RAAF.[107]

Life as a pilot and observer meant a crashed aircraft could dump them in their own lines or behind enemy lines. This latter could lead to being captured by the Turks or by the Bedouin who were likely to sell the pilot to the Turks for gold. Or kill him and loot his clothing and weapons. Neither fate appealed. It thus became common practice for pilots to attempt to land to

---

[107] The Mills, Williams and McNamara stories are related in many sources but these three cited are from Molkentin, pp. 83-9.

pick up downed mates and take them home to have another go the next day.

Gaza, the next major town on the way to Palestine, had to be taken. From the Savoy Hotel in Cairo, Murray would plan the attack from 150 miles away. General Dobell, Commander Eastern Force, would be the Ground Commander after considerable input from Murray, who once again, failed to comprehend that remote command meant nothing more than interference to those who had to get the job done. General Chetwode, Commander Desert Column, would be battle commander to execute the overall attack.

Three lieutenant generals wondered who was in charge of this exactly. Then, at the last planning moment Murray advanced with his HQ to El Arish, still more than twenty miles behind any action or visibility of action, expecting to run the show. Dobell and Chetwode co-located their respective HQs that had two staffs doing one job and getting in one another's way. Disaster, unsurprisingly, was afoot.

11: Gaza: Two Tales of a City

Leadership is an art – victory is a favour of God.
– General Liman von Sanders

Murray wanted to drive the Turks out of Palestine. First he would have to capture Gaza with its well dug-in and heavily protected garrison, supported by machine guns, artillery and reinforcements all around.

But would the Turks hold Gaza or evacuate as they had at El Arish? Murray didn't know if they'd leave or defend with what they had, or even reinforce. And it wasn't just Gaza that had to be captured. There was a southern line from Gaza to Beersheba, 25 miles to the east; this was the bottom of the Turkish-held ground and all would have to be taken for a real advance to be made.

Murray had moved his HQ to El Arish, far enough away to have no visibility of the action but close enough to interfere. The Commander of Eastern Force, Dobell, was looking over the shoulder of the Desert Column Commander, Chetwode, which wasn't too hard as they were side-by-side on the ground. Outside this clique were the various fighting divisional commanders of the British infantry and the mounted divisions, Chauvel and Hodgson. There were now eight generals with layered responsibility.

The missing link was robust leadership at the top. Murray should not have been the commander as he'd already appointed two tactical commanders and that was one too many. He'd achieved his mission to defend Egypt and the Suez Canal and was thought by his friends in the War Office to be doing okay. To the officers around him, it was command chaos. His tactics cost lives, delayed victories and created disharmony. Worse was to come.

Gaza is about two miles from the Mediterranean coast on a 200-foot high plateau, separated from the sea by sand dunes. Around much of its perimeter huge cactus plants formed a natural barrier, just as dangerous to man and beast as the trenches and barbed wire that surrounded it. It was well watered and the Turks had food and supplies for a long defence. To the southwest, a high feature known as Ali Muntar overlooked Gaza. This had to be captured early to protect the main attacking force and make the defenders position in Gaza difficult.

To the south of Gaza was Wadi Ghuzze where water could flow swiftly after any rain. This deep ravine-like cutting could conceal sizeable forces of infantry and mounted troops and the engineers had to dig ramps for the mounteds to ride up.

Intelligence reports, supported by aerial observations, first indicated there were around 2,000 Turks in Gaza. That sneaky von Kressenstein, based on his own aircraft observations, knew the British were coming. He moved in another 2,000 troops plus two Austrian and one German artillery battery. He had up to 10,000 troops scattered in three locations within ten to twenty miles of Gaza and all had water.

Once more, water was the major issue for the British force. Gaza was known to have plenty. Surrounding areas had wells and cistern springs but drawing it up was slow and unsuited to mobile operations involving thousands of men and animals. Gaza's water was the key; take it and take it quickly, or go home.

In the early months of 1917, British intelligence learnt that the Turks and Germans were at loggerheads about what to do. The Turks wanted to withdraw from the Gaza/Beersheba line towards Jerusalem. At the same time British intelligence heard that the Turks were abandoning southern areas where the Arab Revolt was being pursued and would join von Kressenstein in Palestine. The Germans, on the other hand, intended to hold the Gaza/Beersheba line. Murray realised that the Turks had withdrawn from previous fortified positions; he thought they would do so again. He planned to block or chase, rather than a serious attack. Nevertheless, he built up huge stores and supply depots at El Arish.

Murray wrote the War Office in early March 1917:

> The latest information appears to indicate that the enemy retirement will not stop on the Gaza/Beersheba line; further, it would appear that the force at the enemy's disposal is inadequate to hold a line of so great an extent, and, should he hold it, it can be broken by us at any point. It will be some time before the enemy's forces in Arabia can join the force south of Jerusalem.[108]

Murray also believed that his EEF could wrap up the whole of southern Palestine. On 15 March, Murray again wrote that the Turks looked ready to pull out of Gaza before the British advance and that 'the small detachments left at Gaza will not be able to offer any effective opposition to Desert Column.'[109] He was wrong.

To give Murray some strategic credit, he didn't want the Turks disappearing too quickly; he wanted to join them in battle, give them a hiding, destroy their army and advance into Palestine quicker, thus putting an end to the war in this theatre. He believed he had the superior force with three infantry and two mounted divisions. His problem wasn't comparisons of force, however, it was the logistic ability of the British forces to keep up; including the railway and the water pipeline – no army won a war by neglecting logistics. But there was another downside; he was short of artillery and the RFC and AFC were outplayed in aircraft performance. This gave the enemy a distinct aerial superiority that could hamper a ground assault.

Time was needed to prepare. Accumulation of stores and fighting equipment and rest for the men was essential.

---

[108] Gullett, p. 259.
[109] Gullett, p. 259.

Whenever you get a mix of Brits, Aussies, Kiwis, Indians and whoever else pops up, you have competitive spirit. Sport is a good antidote to stress. It was too hot for rugby so the Desert Column held its First Spring Meeting on the old battlefield at Rafa on 21 March. A series of races was planned, a program printed, trophies obtained, an enclosed paddock constructed, a betting ring established, a beer tent found and spectator enclosures for all ranks. Any rank could ride.

It was a remarkable preliminary to the first Battle of Gaza; an event to take men's minds away from the ordeal ahead. Preparations were made with enthusiasm. Chauvel gave a cup for the Anzac Champion Steeplechase, open to all ranks. The Sinai Grand National was won by a horse wounded earlier on that very battlefield and the Jerusalem Scurry was won over five furlongs by 'No Name' for the 2nd Battalion Sikh Pioneers. A total of £200 was raised for charity. At the heart of the event lay the love of the horse shared by Australians, New Zealanders, Indians and Englishmen, and the strong competitive spirit between brigades and divisions. Loads of fun and excitement momentarily broke the boredom, toil and danger of the desert. This also served to informally bring together the diverse men that made up the Desert Column.[110]

Then the first battle of Gaza was on.

The plan was sound enough. Use the British infantry to capture the heights of Ali Muntar. Use the mounted troops to surround the south, east and north to prevent reinforcements assisting the defenders and knock over any defenders trying to escape. Then use infantry to attack from the south and southwest into Gaza and defeat the Turks. If needed, the mounteds could come to the assistance of the infantry and attack from their surrounding positions. A night march on 26 March was needed to create a surprise attack next morning.

One problem though: von Kressenstein had snuck more soldiers into Gaza under the cover of darkness. Now they were now facing 12,000 troops. British intelligence missed this.

---

[110] Hill, pp. 100-1.

Gaza: Two Tales of a City

All were ready to move to their position for the morning assault. Bugger, fog! Although they hadn't anticipated the fog, the horsemen, having trained for night moves in clear weather and fog for months previously, got through Wadi Ghuzze and were in position by the appointed hour. The infantry, however, untrained in night or fog moves, faulted as General Dallas dithered about, wondering where they were. Two hours late, with everyone else chomping at the bit, he got his troops somewhere near their start line. But his artillery was nowhere to be seen. It finally turned up but couldn't range on the enemy as it couldn't see them. Then they were short of big guns and ammunition. They were ineffective in supporting their infantry. Now around five hours late, and having been absent from his HQ while General Chetwode was attempting to get him to start his attack, Dallas finally got 53rd Infantry Division on the go. The poor blighters were caught in sand and heat. They were also caught in the sights of the Turkish artillery and machine guns. Their losses were high.

Due to the delay, Chetwode placed Chauvel in command of both Anzac and Imperial Mounted Divisions with orders to attack the town from their surrounding positions. Many of the Turks had been moved to face the infantry to the south so the mounted task was carried out without difficulty. They entered Gaza and penetrated deep, capturing many prisoners.

The day was ending, dark was falling. By now the 53rd had been supported by the 54th Infantry Division, up till now inexplicably held in reserve by General Dobell rather than given to Chetwode as attack commander. Many of the infantry had got through the cactus, full of nasty snipers and machine guns: more casualties. Dallas and his battalion and company commanders, once they got going, had done a sterling job in the face of heavy Turkish fire. Approaching Ali Muntar,

> They had 4,000 yards to travel against an enemy in a high situation and absolutely concealed in earthworks and cactus hedges; and the many Australians and New Zealanders of the mounted regiments who watched

the attack appreciated for the first time in the war the splendid steadiness of the British infantry and the fine quality of its regimental leadership under the most galling conditions.[111]

Unfortunately, their artillery fired over the top into a cemetery unoccupied by living beings:

> But the battalions, changing their tactics to a series of rushes, and very gallantly led by platoon and company officers, struggled gamely on under deadly punishment.[112]

The missing British artillery and Dallas's delay, even after the fog had lifted, has never been explained, either by Dallas or any of the three commanding generals' in their 'after action' reports; no surprise really, when such collusion would protect them from cries of incompetence. Nevertheless, at twilight the infantry were on Ali Muntar and inside the perimeter and were able to hold ground. The mounteds were well established in the back of the town and ready to accept a major Turkish surrender, white flags having been widely distributed. Things looked good. The divisional and brigade commanders doing the fighting were pleased with their accomplishments.

On the same day, early that morning, a Turkish general had been riding in his buggy towards Gaza to take command of its garrison, unaware of the British advance. He was captured by a small group of Australian horsemen. Terrified by stories the Germans had told of Australian treatment of prisoners, the general tried to be courteous and took a gold case from his pocket and offered the horsemen a cigarette. Unimpressed, one of the lads removed a half smoked, fully squashed fag and offered the return courtesy to the general who ungraciously declined. Chauvel was most pleased with this capture. The general, however, was most offended when he demanded an officer of equal rank to escort him to POW camp. Chauvel

---

[111] Gullett, p. 275.

[112] Gullett, p. 275.

Gaza: Two Tales of a City

replied that he was the only divisional commander and had other things to do, so he allocated a junior officer to escort duties and dispatched a very indignant general towards Murray.

The horsemen's spirit was almost playful. Their day hadn't been too difficult, given the nature of their tasks and their horsemanship. At one stage the 5th Regiment was advancing at great pace despite the cactus hedges:

> The Turks, said one of the officers later, ran in and out like rabbits, and we shot them as they ran. One old farrier-sergeant, who had joined in the charge, was finally cornered in a field with impassable cactus; and while his horse nibbled at the grass, he continued sniping over the hedges from the saddle. 'Why don't you get off?' a passing officer asked him. 'I can see better from up here, he answered and went on shooting.[113]

At this stage, Murray really had little idea and abandoned things to Dobell. Dobell had been told Turkish reinforcements were closing on Gaza. Dickie Williams, however, records that, while he was on air patrol at 4.30pm

> It appeared that darkness might not only allow the infantry to rush the Turkish position, but it would probably find the mounted troops in Gaza itself and behind the Turkish trenches which were holding up the infantry ... Williams noted that some Turkish infantry were approaching on foot from the north, but concluded that the mounted troops and armoured cars covering the encirclement could easily deal with them. Williams landed at Dobell's command tent with this positive news. Since he had been there in the morning however, the atmosphere had changed. He found the Chief Intelligence Officer (CIO) sitting there quite despondent.[114]

In a strange conversation, Williams passed this information to the CIO who replied 'Oh, very interesting'. No more was said. Williams went back to his base, fully expecting that by

---

[113] Gullett, p. 279.

[114] Molkentin, p. 98.

dark the British would have Gaza.

Neither Dobell nor Chetwode seemed to have much understanding of the strong position of their troops. Perhaps their staffs hadn't passed on the vital information. Perhaps they were overly concerned with water and reinforcements that weren't actually coming yet. While Dobell understood the need to refresh horses and men, he didn't realise that many of the horses had been watered throughout the day. He was unaware of the ground taken by 53rd Infantry Division, primarily because he didn't leave his HQ even though it was less than three miles from the action. He could have ridden or used a light car. His erroneous thinking was that the regiments of the Anzac mounted were scattered throughout Gaza yet many were now dismounted and the garrison nearly subdued. His major concern was an imminent approach of Turkish reinforcements that could counterattack the next morning when his force was unprepared for defence. But he didn't realise that those numbers were an overestimate. He was unaware of Williams report. His communications seem non-existent. But his decision was made:

> Chetwode telephoned Chauvel about 6pm. He described the situation as he saw it and told him that he had decided to withdraw the mounted divisions. Chauvel, in view of his grip on Gaza, protested against a withdrawal.[115]

Chetwode and Dobell had conferred and agreed. There had been no consultation with the tactical commanders. They either didn't receive or didn't assess information from the field. Both had failed to go forward and view the tactical situation, preferring to sit in their HQ, using a telephone and notepad to communicate with forward commanders, whose opinions they seemed to have ignored anyway. Therefore, neither really understood the situation of the infantry or mounteds.

When Dickie Williams was told later that night that the British were withdrawing he was overcome, 'Retiring? I don't

---

[115] Hill, p. 108.

believe it'.

It was left to Chauvel to order his light horsemen to withdraw. The brigade commanders now occupying the town were dumbfounded. They had

> Made great advance into the very base of the enemy's strength, had found the Turks demoralised and disinclined to fight, and had suffered practically no casualties. Every indication pointed to a rout and general surrender at any moment. Chaytor expressed his opinion of the order by exercising his right to have it in writing before acting upon it. Ryrie bluntly told his staff that there was to be no withdrawal until every trooper had been collected. Not a man is to be left behind. As the order was slowly communicated from brigade to regiment, to distant squadrons and troops, it was everywhere received first with doubt and then with disgust. Again and again the astonished and puzzled officers ordered their signallers to have it repeated; and, when its truth was beyond question, they felt as men could only feel who were ordered to accept defeat, when in their opinion the battle was won and the objective actually in their hands.[116]

Even von Sanders couldn't comprehend the British decision based on his knowledge of the Turkish position:

> With the customary Turkish delay in starting, the march for the relief of Gaza was delayed into the afternoon hours. Their action did not become effective on 26th, as both [relief] columns were several times checked during the march and it was 9am on 27th before they were near enough to Gaza for the relief to become sensible.[117]

The ride back across Wadi Ghuzze was gut-wrenching. Mates had died and been wounded, but now their effort was absolutely wasted for no good reason. Men who'd been kept alert by the exhilaration of the fight, albeit, going for three days without sleep, were now so flat they were once more falling off horses with exhaustion. The infantry were kept close and by

---

[116] Gullett, p. 283.

[117] Von Sanders, p. 165.

next day's sunrise, ordered to once more attack and take Ali Muntar. The Turks, now heavily reinforced after the mounted forces had withdrawn, were never going to let this happen.

Confusion reigned everywhere. Even the Turks thought their case was doomed. Intercepted Turkish wireless messages on the night of the attack confirmed the views of Chauvel, Chaytor and Ryrie. From von Kressenstein to the Turkish commander, Tala Bey, 'Having regard to the disposition of Turkish troops, can an attack be successful at early dawn? I beg you to do your utmost to hold out so long'. To which Tala Bey replied, 'Your telegram received. Please attack at all costs at 2 o'clock tonight'. At the same time Bey sent this to his own HQ: 'Position lost at 7.45pm'.[118] It seems these interceptions weren't passed in a timely manner to Chetwode, or Dobell or Murray or anyone else who thought they were running this show.

To withdraw was a tactical failure. The hard-won victory of the infantry and mounted troops was instantly turned into a defeat. Disinclined to fight in the dark, the Turks weren't aware of the night withdrawal of the British. It wasn't until morning when Tala Bey fell about laughing and laughing and laughing. This was a gift from Allah:

> The fundamental failure at Gaza was not one of intelligence nor even staff work, bad as some of it was; it was a failure in command. Like General Lawrence at Romani, neither Dobell nor Chetwode made any attempt to go forward to speed up the sluggish 53rd Division or to confer with Hodgson when the Turkish pressure came on ... instead, both elected to conduct the battle like a training exercise for junior officers. Horses and even motorcars were available and there was good going for both.[119]

A well-formed plan disintegrated through ineffective triplication of command. Murray blamed the fog for the delay and Dobell for not using all his resources. Dobell blamed Dallas for not moving quickly. Chetwode was hamstrung by

---

[118] Gullett, p. 289.

[119] Hill, p. 106.

Dobell's clinging to an infantry division depriving him of all the infantry. Chetwode didn't go forward.

Chauvel showed his disappointment when he uncharacteristically wrote to Birdwood a few months after:

> There were two reasons it failed. The chief was that the infantry treated the fog as an obstacle and waited until it had cleared instead of looking upon it as a God-send and making every use of it as we did. The other was that the whole force was not put in.[120]

Like a scene out of *Dad's Army*, on 28 March, General Murray sent the following curious message to describe the Gaza action to the War Office:

> We have advanced our troops a distance of 15 miles from Rafa to Wadi Ghuzze to cover the construction of the railway. On 26th and 27th we were heavily engaged east of Gaza with a force of about 20,000 enemy. We inflicted very heavy losses on him. It is estimated that his losses were between 6,000 and 7,000. We have an additional 900 prisoners, including the commanding general and the whole staff of the 53rd Turkish Division'.[121]

But that's nothing. In his report, General Dobell reported:

> This action has the result of bringing the enemy to battle, and he will now undoubtedly stand with all his available force in order to fight us when we are prepared to attack. It has also given our troops an opportunity of displaying the splendid fighting qualities they possess. So far as all ranks of the troops engaged were concerned, it was a brilliant victory, and had the early part of the day been normal victory would have been secured.

'Thus do men deceive themselves'.[122]

Funny how the Turks saw it differently from these two generals. 'No longer was there talk of withdrawing to the Jaffa-

---

[120] Hill, p. 107.
[121] Keogh, p. 102.
[122] Keogh, p. 102.

Jerusalem line. They would fight and win where they stood'.[123] And:

> Now, with his troops elated, and with a very low opinion of the quality of the leadership opposed to him, he would have no hesitation about standing on the strong Gaza-Beersheba line'.[124]

This is one more illustration of Professor Dixon's view of the military incompetence of certain generals of their day, on whom the lives of men depended.

And so ended Gaza I.

However, a second would be worse than the first.

General Dallas was relieved command of the British 53rd Infantry Division. Generals Murray, Dobell and Chetwode remained.

Professor Dixon pointed out a funny thing about human nature. Some people, when confronted with their errors, especially people with authoritarian attitudes, tend to point the finger at someone else. Then, some even go further in attempting to overcome or support their previous errors:

> Both Dobell and Murray were clearly determined to put the best possible complexion upon the engagements of March 26th while they applied themselves to preparing a second attack with all the resources at their command. They aimed at achieving a decisive victory over the Turks in a pitched battle, and by this triumph to smother up the fiasco of March 26th. During this time Dobell repeatedly advised the Commander-in-Chief that the outlook was exceptionally bright, and Murray unfortunately appears to have accepted these assurances without question.[125]

Around this time, things were still not going well for the British in France and Salonika and home morale was declining. A success was desperately needed, so liberating Jerusalem and the Holy Land continued to be thought desirable by the

---

[123] Keogh, p. 111.

[124] Gullett, p. 295.

[125] Gullett, p. 297.

politicians in London. Preparations were made over the next three weeks, while Gaza held out.

Of course, this also gave the Turks, now bulging with confidence, the same time to improve defences. From Gaza to Beersheba and the surrounding hills and plateaux, reinforcements, trenches, sandbags and barbed wire were all strengthened. The Turks dug in with more soldiers, artillery and machine guns.

General von Sanders said before the German/Turkish advance on Romani a year earlier, 'Anyone who has had the responsibility in a difficult situation knows that in war it is sometimes necessary to carry out hopeless undertakings'.[126] Gaza II was very much a hopeless situation for the British from the start.

Although, this time the British had secret weapons. Six tanks and 2,000 poison gas canisters. Murray and Dobell made much of these before the attack. Unfortunately, for them, neither was a success in the soft sand and desert winds.

The complexities of the battle make poor writing and horrid reading so summaries will suffice. Once more, the plan was simple, albeit tactically ridiculous. A full frontal assault by infantry and dismounted horsemen and cameleers. Oh, and use the AMD to chase the Turks when they ran for it or to prevent reinforcements from arriving.

The issue became that the Turks, having seen the quality of British leadership in Gaza I, didn't run for it: they stayed. They ranged their artillery and machine guns, used the German aircraft to detect then smash ground troops, and they had well-prepared positions. The combined British artillery and naval gunfire was insufficient to bombard the Turkish lines and take out their guns. The British foot soldiers had to travel over open ground with the cornfields and red poppies of the plains around the town as their only cover, uphill, in the heat without adequate air or artillery. The Imperial Mounted Division with its two Anzac brigades dismounted and attacked infantry-style.

---

[126] Von Sanders, p. 143.

Desert Anzacs

The AMD had little to do in this engagement in any chasing role; there was no cause for chasing.

Divisional commanders and brigade commanders from the beginning had doubted the likelihood of success. The soldiers just knew it was crazy. Their fears were correct. It took 6,000 casualties from 18-20 April 1917 for the British to be defeated and Dobell to call it off. However, the bravery and dedication of all ranks in the British force was a thing to behold; a major reason for the high casualties as they just wouldn't stop.

A few events worthy of note show the character and mateship that Anzacs hold dear. The medics had established a tent dressing station as far forward as they could. But as the Turks advanced it was too close:

> Our troops started to retire right up to our dressing tent. We were practically in the front line, with machine gun bullets whipping up the dust. It was obvious we had to retire in a hurry. Walking wounded and a few on stretchers were moved back. But we had run out of stretchers and sand carts. Then the answer came to me in a flash right out of the blue. Something not in the textbooks, and never before mentioned, tried or thought of. Dozens of discarded rifles and bandoliers were lying around. By putting three bandoliers, at head, centre and foot, around two rifles, held apart by stretcher supports, we improvised stretchers. Then calling for willing helpers from retreating troopers, we loaded all remaining wounded onto these emergency stretchers and retreated slowly. The Dressing Tent and other gear was captured shortly afterwards.[127]

The aircraft of the RFC and AFC, as well as those of the Germans, flew overhead each day. The bombing was furious. Strafing hit men and horses. The superior German aeroplanes were better able to damage British and Anzac lines and as the defenders had no anti-aircraft guns, the German air attack was devastating. At one time a fight between a British and German aircraft had the ground troops cheering as one wing fell off and the plane dropped like a stone giving the pilot a violent end.

---

[127] Hamilton, p. 23.

When they recognised the red, white and blue tail marks, the cheering abruptly stopped and the shooting recommenced.

The stretcher-bearers showed no fear throughout the battle. Working in the open there was no hope the enemy would acknowledge and respect their humane mission, yet they hung in, despite their own casualties. Captain W. Evans and four bearers worked all day under fire. This tiny group treated 240 wounded. Private Tibby Cotter, the Australian test cricket fast bowler, was prominent in attending the wounded for three days.

By the afternoon of 20 April, General Hodgson's Imperial Mounted Division had been pulled back with orders to dig in and prepare for a counterattack 'This was the fourth night on which the horsemen had been without sleep, yet digging was carried on at high pressure'.[128] The counterattack didn't come and it was only afterwards that the Germans indicated that a counterattack could not take place due to their shortage of ammunition. Hodgson and the division were saved an almighty battle that could have gone either way.

The War Office now took a dim view of Murray's early confidence, followed by his total failure to achieve any objectives and the high number of casualties. The battle was seen as Gallipoli II.

Of the Germans, Colonel Keogh reported:

> In his account of the battle von Kressenstein states that at no stage was he seriously pressed and that his troops were not much troubled by the British artillery, including the fire of the ships' guns. Before the battle began he felt some concern that the British might, despite indications to the contrary, put in a strong attack at Beersheba where he could afford to only station a very small garrison. When the British attacks ceased his troops were so unshaken that he thought of launching a general counter-attack with a view to driving his opponents back across the Ghuzze, a course he was reluctantly forced to abandon because of shortage of ammunition and transport. From his point of view the

---

[128] Gullett, p. 329.

battle was a 'walk-over' from start to finish.[129]

On 21 April, Murray relieved Dobell of his command. Chetwode was appointed Eastern Force Commander and Chauvel Desert Column Commander. Chauvel was the first Australian and first dominion officer to command a corps; a massive milestone in British military records. Chaytor, the New Zealander, was promoted to major general to command the Anzac Mounted Division. Speculation now erupted in his own EEF, in Cairo, France and London that Murray's neck was 'on the block'. Nearly two months later, on 11 June, he was advised that he would be recalled to England and replaced by General Sir Edmund Allenby. He took it graciously although he thought it was unfair. No one else seemed to mind.

Chetwode quickly realised that the Turks were well fortified in and around Gaza and were building their defences from Gaza out towards Beersheba. But the further away they moved the weaker the strength of their defences. There was thought to be little water between Gaza and Beersheba. He also recognised that if the British Government was fair dinkum about taking Palestine and Jerusalem, it would have to give it the priority it was giving to France and the Western Front – now there was a challenge for politicians and militarists sitting in London. Their first reaction, however, was to take away 60,000 British infantry with battle experience and replace them with inexperienced but enthusiastic Indians.

Summer was now in full swing – stinking hot. Chetwode, unencumbered by Dobell, came to the fore. Concerned about the possibility of a Turkish attack, he quickly had his troops digging their own defensive trenches. In addition, they had to conduct aggressive air, car and mounted patrolling to keep the Turks honest while training all troops, especially the new Indians. He devised a rotational rest cycle for the exhausted soldiers and horses to regain their condition and stamina. Sending blocks of men to Egypt or the Mediterranean, swimming, playing sports and plain old lying around, worked

---

[129] Keogh, p.118.

wonders. The rotation of the mounted brigades and infantry battalions, together with Chetwode's energy and cheerfulness, brought a whole new energy and a spirit of high morale to the troops. Thus, he lifted them from the doldrums they had fallen into under Dobell and Murray.

Nevertheless, this was no summer vacation. Soldiers had no choice but to mix whatever rest they could get with the intensity of day and night rides in heat and khamsin dust storms. Dust settled on lips, under eyelids and up nostrils, where lumps grew like fur balls and ear canals filled like sandbags. And the flies loitered while seeking out the tiny cuts and scratches they could convert into septic sores and infections. The less than exciting diet of tinned beef and tea, was varied by tea and tinned beef.

There were no pleasant villages with nice people to visit. Chauvel had, however, encouraged the Red Cross to erect their own rest camps where men could escape army routine for a few days. They could swim and treat their sores, scratches and itches. Read, write, do nothing. At one time, a medical officer concluded that one in three of the troopers was suffering from dilation of the heart. Chauvel's rest camps helped many. But amazingly, certain dimwitted staff officers from Murray's HQ complained about the rest camps. Bless Chetwode; for he didn't speak to the offending staff officers. Instead he messaged direct to Murray, 'It is my business what Desert Column does, no one else's'.[130] And that was that.

With all these goings on, Chetwode never lost sight of the strategy. He was a leader inclined to action. Patrolling and training were necessary and to his delight, 'The British Government, for the first time in this campaign, made it clear that it was whole-heartedly in the prosecution of Palestine'.[131]

Chauvel was called in by his now good mate Chetwode and shown a plan of attack that his Chief of Staff, Brigadier Guy Dawnay had developed. Murray, still around but with

---
[130] Hill, p. 115.
[131] Gullett, p. 342.

departure imminent, was left out of the discussion. Gaza had to be taken, but not in the way that had been twice tried.

Not only had Gaza town been strengthened, so had more of the surrounding hills and ridges. Trenches had been dug but without barbed wire, either through a lack of supplies or insufficient transport to get it there, Turkish logistics once more their downfall. Chetwode knew a full frontal attack would be suicide and the men really weren't up for that again. Further east, though, the defences around Beersheba were less robust. If Beersheba was taken, they could move towards Gaza from the side where defences were also less robust, put out a screen force to prevent reinforcements charging onto them and belt Gaza from the rear and sides. But it was a long ride for men and horses in a waterless area that involved a near 50-mile ride to sneak up without being spotted by the Turks and German aircraft.

Once more water, the key to desert operations, was a major consideration. The plan Dawnay and Chetwode presented to Chauvel was:

> Based on deception, speed and surprise ... a feint at Gaza followed quickly by a powerful thrust at Beersheba with Chauvel's mounted troops. With Beersheba and its water in his hands Chetwode could launch his main blow against the enemy's left and prevent withdrawal of their force or reinforcements with the Desert Column. Speed in the initial operations would be vital because except for the wells of Esani (many miles away) all the known water was in Turkish hands; Beersheba must fall in the day, and so quickly that the enemy would be unable to destroy the wells.[132]

To the horsemen, lack of water was more formidable than the Turks. But Chauvel had a smart engineer officer, British in fact. Brigadier Russell knew that two ancient villages were thirteen and fifteen miles from Beersheba and both previously at least, had good water supply. If the Desert Column could circle via these they would be out of sight of the Turks and

---
[132] Hill, p. 115.

could hydrate effectively. As it happened, there was a Turkish railway line not too far from those old villages. Chauvel set about a diversionary attack on it while confirming that water was still available in good supply. Both activities were positive. The planning was on.

Now, in early June while all this was going on, Murray was still in the chair and no one was certain when he was going and who was coming. This was not a plan to present to him so life continued as if there was no plan, until the announcement: Murray going, Allenby coming. Hold the plan for Allenby.

In the meantime, Chauvel was the Australian Commander from Cairo to the front line. His command included hospitals, remount units, the veterinary hospital, rest camps, the training of the yeomanry division now in his corps, admin HQ in Cairo, and training programs throughout the corps. To maintain this vigorous workload, he swapped his horse for the faster motorcar. His own health, now under the watchful eye of the concerned medical director Colonel Rupert Downes and the letters Chauvel constantly wrote to his wife, also occupied this energetic man. 'He almost lived in his Ford car, which, whatever it lacked in comfort, made up for in robustness and cross-country performance'.[133]

Down in the Hejaz, the Arabs had come to life.

---

[133] Hill. A.J., p. 116.

12: The Arabs Have Aqaba

Arabs may not tell the truth but they will believe what they say.
– T.E. Lawrence

In May 1917, the desert launched into summer with 50°C heat from a flaming sun – every day. The Arab Revolt had stalled in the Hejaz. Yet the Arab armies and some of the tribesmen had received some newer rifles to replace the outdated Japanese ones they'd received earlier. They'd received British officers for demolitions and instruction and scratchings of Egyptian artillery. They were supported by ship borne aircraft of the RNAS, launched from *HMS Ben-my-Chree* and *HMS Raven* and three more warships for coastal attacks and troop transport. Included in this armada was the on-loan Australian steamer that had been converted to *HMS Suva*. General Murray had been disinclined to offer his limited land resources to Sharif Hussein, as an untested tribal leader. The Arab forces had captured coastal towns along the Red Sea and were doing minor damage to the Hejaz Railway. Medina and Ma'an were still in Turkish hands and things were not progressing. Besides, the British, and especially the French, didn't really want the Arabs progressing too far north towards Syria for political reasons. Nor did they want them to get ahead of their own forces in Palestine. The pressure was, therefore, on the Arabs to prove their revolt worthwhile.

Aqaba, the little port town at the head of the Gulf of Aqaba (the Gulf of Eilat on some modern maps), was vital to the longer-term plans of the British, the French, the Arabs and the Turks, all for different reasons. The Turks held it; others wanted it.

The Arabs Have Aqaba

With its scanty mud houses, crushed coral beach, wavering palm trees and clear waters, Aqaba contained a small stone fort (originally used as security and a water source for Egyptian pilgrims on their *Hajj* to Mecca) that held a detachment of 300 soldiers supported by artillery. It's the only seaport for Transjordan and southern Palestine. On the eastern side of the gulf are steep mountains running down to the water, while to the west is the Sinai Peninsula with its steep mountains. Heading north and landward out of the port is Wadi Yutm, that from its southern end is very narrow with near vertical mountains both sides and peaks in excess of 5,000 feet. At its start is Khadra, out of sight of ships and out of naval gunfire range, with trenches able to hold 300 soldiers. The wadi initially runs east-west some six to eight miles before turning north-south, where a fort of Roman origin, named Kithara, could hold 150 soldiers. Continuing north through the wadi another 25 miles was the little fort of Guweira holding a garrison of around 100 soldiers.

Aqaba gave access northwest through Wadi Araba and the Negev Desert towards Palestine; southwest into the Sinai and Egypt; northeast towards the rich pastoral lands of grain that fed the Turkish forces; east towards the Ma'an garrison and the Hejaz Railway; southeast towards the major transport corridor to Medina. Its capture would be a major blow to the Turks and a major boost to the Arab and British resupplies. It was a solid gold trophy.

The Royal Navy controlled the Red Sea and the Gulf. The Turks only had overland access to Aqaba from the railway at Ma'an to resupply the fort. The French and British suspected the Germans could move shipping mines and submarine parts via the Hejaz Railway into the Red Sea to block the Royal Navy.

Early British and French thinking had been to land troops on the beach supported by naval fire in order to capture Aqaba. This had several purposes: to prevent German mines and submarine parts finding their way into the Red Sea; to fully control the Red Sea; to provide a base for future attacks against the Hejaz Railway and the garrison at Ma'an; and to provide materiel support to its officers Lieutenant Colonel Newcombe and Lieutenant Hornby, already there with Arab tribesmen who'd been conducting guerrilla raids on the railway.

The port facilities could support the Arab Revolt with stores and equipment keeping pressure on the Ma'an and Medina garrisons. It would provide a supply base to the right flank of the EEF for any future move northwards through Palestine to Syria. The British wanted to take it before the Arabs got there and to keep the Arab forces in the southern area. This would reduce the likelihood of the Arabs using it as a steppingstone to Damascus and interfere with the political desire for later French control of Syria. Arab desires differed.

Arab thinking was to capture Aqaba and beat the French and British to it. Aqaba's capture would give them domination over Ma'an and points north towards Damascus and south towards Medina. Moreover, it would indicate to other Arabs, especially those in and around Damascus, that the revolt was both genuine and successful and show the British that the Arabs were a serious fighting force. It would then be justifiable for the British to provide more money, weapons, advisers and food to the Arab armies and tribes through Aqaba than using the existing hazardous overland route from the distant southern town of Wejh, hundreds of miles away.

Down in the Hejaz on 17 May, in the Red Sea port town of Jeddah, Emir Feisal met Lawrence (of Arabia) and Lieutenant Colonel Wilson, the British military adviser. Also there was Sheik Auda abu Tay, renowned as a ferocious warrior who had killed more unworthy tribesmen and Ottoman soldiers than the uncountable number of children he had reportedly sired. Auda's reputation was godlike, as was his supposed ability to recruit other tribes. He was the leader of one of the biggest and fiercest tribes, the Howeitat. But tribal loyalty was never guaranteed in the desert. To show his loyalty during their meeting, Auda leapt to his feet, bolted through the tent flap and was heard furiously banging and yelping. Investigation found him hammering to powder with a stone, a set of false teeth that had been given him by the Turks, screeching 'God forbid, I am eating my Lord's bread with Turkish teeth!'[134] Convinced by Auda's display of sincerity, the meeting with Feisal resumed. A plan to capture Aqaba was hatched.

Auda, incentivised by the offer of £20,000 in British gold to share around, was adamant that Aqaba could be taken by tribal Arabs as this was his tribal turf. But they would have to negotiate one of the worst deserts on earth.

> The force would have to pass through enemy lines and travel hundreds of miles on camels through desert just to arrive at Auda's camps near Ma'an. Once they arrived at his camps they would have to raise an army in the shadow of the Ottoman's regional HQ and to conduct feints to keep them from suspecting the real plan of attacking Aqaba, still nearly 100 miles away.[135]

It was agreed. Auda would be a dominant person but the overall commander would be Sharif Nasir, already commanding the Northern Arab Army (NAA). Feisal had the monstrous task to produce some sort of tribal unity to maintain the momentum towards Damascus.

Unlike the story told in the epic movie or his acclaimed *Seven Pillars of Wisdom*, it seems Lawrence was not involved in the planning or the attack on Aqaba, and he sure didn't lead it. He was a junior officer with little battle experience at this time. It seems he expressed a desire to attend, pointing out his demolition expertise, and was thus invited.

Tribal sensitivities now cascaded.

> Feisal could not simply send his army northwards to take Aqaba. The Bedouin clans and tribes had their own territories and practised a system of feuding, raiding and war for booty. It was necessary to obtain permission for

---

[134] Lawrence, T.E., p. 228.

[135] Scott, J., *Conflict Archaeology in Southern Jordan, Wadi Yutm and the Arab Revolt 1916–1918*, p. 82.

use of Bedouin wells and grazing. To organise tribes for war, feuds often had to be settled or agreements reached before they could unite.[136]

Maintaining tribal Arab support is complex. However, promises of gold and weapons, food and explosives, go a long way. Fulfilling promises was the task allocated Wilson and Lawrence.

The only English eyewitness account of the capture of Aqaba is Lawrence's *Seven Pillars of Wisdom*, written after the war based on his memory plus, his official reports to the Arab Bureau in Cairo during the war. Strangely, there is often disparity of various events between Seven Pillars and his official reports. Nevertheless, John Scott, a PhD conflict archaeology student, and whom the author accompanied, conducted three field archaeology expeditions in 2010, 2011 and 2013. During these expeditions we examined the battle sites on approach to Aqaba and determined that Lawrence's descriptions were supported by what remains on the ground today. It's just his telling of his level of personal involvement that is disputed.

During those expeditions, we sat on the desert floor several times talking to local Bedouin. We shared mugs of sweet mint tea and their version of damper cooked in coals, then torn apart and shared around while an elder played his *rababa* (single string device like a bass played with a bow) to songs and recitals of tribal history in poem. At bygone desert dinners, stories have been passed down by generations that told the warriors' tales. 'My grandfather fought many times with Auda and Aurens (Lawrence) and would be given a gold sovereign for every captured Turk and then allowed to loot the ones killed'.[137] At a nearby village where the Bedouin had been relocated from their nomadic tents to shacks of brick and tin, we met the village sheik:

---

[136] Scott, p. 85.

[137] This story was told by the 62 year-old Mohammed Suleiman El Gidman, whose grandfather was one of the tribal warriors with Auda, in an interview with John Scott, the author and a translator from the Jordanian Dept of Antiquities during our field trip in 2011 and is related in Scott's Dissertation, p. A.2.

Figure 15: Sheik Salami Saleem, relating stories to the author.

Figure 16: Mohammed Suleiman El Gidman plays and sings for the author.

Desert Anzacs

> At the last battle in Wadi Yutm with Lawrence and Auda, they offered money for every Turk captured alive 'so they could take them to London'. Some of the Turkish soldiers had deserted and lived with the Bedouin.[138]

Sharif Nasir and Auda gathered their warriors and, accompanied by Lawrence, rode for many days through one of the world's hottest and most waterless deserts, undetected by the Turks. On arriving at Guweira they discovered the local sub-tribe of the Howeitat had subdued the Turkish garrison and were proudly sitting in the fort with trophies of weapons and 100 despondent prisoners.

Next was Kithara. The tribesmen snuck up on the fort but it was bright moonlight and detection was imminent. The academic Lawrence, as recorded in *Seven Pillars of Wisdom*, held them in check, waiting. He knew there was a very convenient total eclipse of the moon just moments away. When the eclipse occurred, the tribesmen jumped from the desert floor, then screamed and howled like ungodly beasts while their rifles were fired to such effect that no one was shot. They pounced and quickly captured the fort with the superstitious Turks throwing down their weapons, terror-struck at the screaming and the sudden satanic darkness. During our research we checked, and indeed there was a total eclipse that night.

Following Wadi Yutm they overran a few outposts before arriving at Khadra. By now there were over 1,000 eager tribesmen busting for the fight and loot. The 300 Turks sensed that surrender was better than mutilation, so they did. This was the whole Aqaba garrison; the fort was empty. There was no further fighting.

On 6 July 1917, Nasir, Auda and the tribesmen strolled into Aqaba and raised the Arab Revolt flag. Pretty simple after all that desert riding and heat. Aqaba port was now under control of the Arabs, ahead of the British and French. Despite the simplicity of the entry, there were wild celebrations by the

---

[138] This story was told by the 100-year-old, almost blind, Sheik Salami Saleem who was a young boy at the time, in an interview with John Scott, the author and our translator, p. A.2.

thousand excited warriors and shots were fired wildly into the air, endangering themselves and the wildlife underneath.

After the capture of Aqaba, Lawrence and a small party took off on camels into the Sinai desert and mountains, heading for Cairo with the good news. They arrived in EEF HQ to find General Murray gone and General Allenby in command. Lawrence was keen to tell anyone who would listen what a splendid job the Arabs had done, how worthy of recognition they were in their Revolt, and how they should now be supported with additional weapons, money, food and military advisers.

Allenby arrived in Cairo less than two weeks before Lawrence. Fresh from his Sinai crossing, Lawrence was still draped in his Arab robes, headdress, golden dagger and bare feet. He found Colonel Clayton, his old boss in military intelligence, telling him the Arabs would be a great supporting force in the east. Clayton took him, bare footed, to Allenby.

Undistracted by Lawrence's attire, Allenby agreed. He invited Emir Feisal and the NAA to become part of the EEF. Feisal could be a lieutenant general to serve on the EEF's right flank and protect the British force. He would be tasked with beating up on the Turks along the Hejaz Railway while keeping the large Turkish garrisons bottled up in Medina and Ma'an. The invitation was passed to Sharif Hussein. Seeing it as their route to Damascus, he accepted.

Two days later, *HMS Euryalus*, the flagship of the Royal Navy's Red Sea Patrol (RSP), anchored at Aqaba to off-load preliminary stores and supplies. By 13 July, just a week after the capture, the incentives for Arab loyalty – money, weapons and food – were being unloaded from *HMS Dufferin* and other ships of the RSP. Getting their attention came shiploads of immediate rewards for the Arabs; £200,000 in gold, 20,000 rifles, twenty Lewis machine guns, eight Stokes mortars, 50 tons of gun-cotton explosives, aircraft, a squadron of armoured cars, food for the NAA for three months, and British officers to provide direction to the Arab warriors.

Further, Aqaba gave a supply base for Allenby's right flank and the Anzac horsemen, cameleers, light cars and airmen. Feisal's army commander, Ja'afr al-Askari and his troops arrived by ship from their southern base on 18 August while Feisal himself arrived on 24 August, aboard *HMS Hardinge*. Soon after, Captain Pisani of the French artillery arrived with guns and Muslim gunners from Algeria. The RFC landed a flight of aircraft at Quantilla around 60 miles to the west. Sharif Nasir arrived with hordes of Bedu tribesmen, encouraged by the gold payments promised them. More French officers, artillery and stores arrived under the command of their politico-militarist Colonel Bremond.

Lieutenant Colonel Pierce Joyce became the Commander of Operation Hedgehog, the British military mission based at Aqaba to support the Arab Revolt east of the line of Aqaba, the Dead Sea and the Jordan River towards Damascus. Hedgehog provided more than 50 British officers as advisers, trainers and demolition experts, and liaison between the EEF HQ and the NAA. The Arab Army would henceforth be part of the action and encouraged to drive north towards Damascus. That got the attention of Hussein and Feisal – Damascus!

As important as military supplies were, gold became the primary incentive for tribal leaders to join or remain with Sharif Hussein's revolt. The tribesmen enjoyed some of this spoil shared out by their sheiks, but plundering and looting casualties in the battlefield came naturally.

Only gold sovereigns were currency, not bullion or nuggets. England had a mint but no mines and a limited gold reserve from pre-war days. India was a big consumer of gold but had neither mines nor a mint. South Africa had mines but no mint. Only Australia had both gold mines and mints in Sydney, Melbourne and Perth.

Private S.C. Rolls (who drove Lawrence's armoured car for eighteen months) reported:

> On the day of our arrival in Guweira Lawrence decided

to make his first raid by car on the Turkish railway. We loaded two tenders with a large supply of gun-cotton, a week's rations and water, a case of 5,000 sovereigns marked Commonwealth Bank of Australia, several coils of electric cable, a battery exploder and several other articles which had been found useful on previous demolition raids.[139]

Lieutenant Colonel George Langley, CO 1st Anzac Camel Battalion, reported:

Cases of gold coins were regularly brought out from Australia and paid out to the Arab Bureau in Cairo for the Arabs.[140]

These sovereigns came from the Royal Mint in Perth via the Commonwealth Bank of Australia, which was the only authority able to export gold in the period of prohibition of gold export during the war years. At that time, the Australian Government wholly owned the bank.

But danger lurked. The Indian Ocean, sloshing on the shores of German colonies in the African continent, was awash with German submarines. Allied shipping plying between Australasia, India and the Red Sea in transit to Egypt then Europe was vulnerable. On Australian wharves and in bars and cafes, spies waited, eager for news of war targets to pass on to their German paymasters. Security demanded silence: 'loose lips sink ships' became the catch cry.

At night, the normally office-bound men of the Commonwealth Bank would throw off their suits and don overalls and darken their faces. Down the Swan River in starlight, they moved the gold by lighter to load aboard vessels already at sea, rather than have waterside workers handle such treasure and heighten the possibility that this information would be passed to enemy agents.

The Commonwealth Bank archival history states, 'Ships would then only sail on assistance and advice afforded by the

---

[139] Rolls, S.C., *Steel Chariots in the Desert*, p. 143.

[140] Langley and Langley, p. 83.

Navy and in no case was any loss occasioned'.[141] Astonishingly, around £20,000,000 was so moved without loss (about 800,000,000 Aussie dollars today).

No one seems to know exactly how many gold sovereigns were handed out but the number is estimated to be many millions, as the Arab armies are on record as receiving 220,000 of them per month for over two years plus a lot of incidentals. Even more passed to tribal Bedouin for per capita captures or shootings. No one knows where they are today, despite major shifting of sands and rocks over the years by local inhabitants, still searching for it. But gold bought loyalty.

Now Aqaba was in British hands and the right flank was protected without Allenby having to allocate scarce British troops to do so. The Arabs were happy, as they were on their way to Damascus and nationhood, or so they thought.

A few months after the capture of Aqaba, Lawrence decided he would have a go at repeating the escapades of Newcombe and Hornby. He'd use machine guns and mortars to give his Arabs even greater confidence. A few enquiries and requests and Lawrence had what he needed: weapons and instructors.

The Lewis gun instructor was an Australian sergeant named Charles Yells, a weapons instructor at the Imperial School of Instruction, Zeitoun, Egypt, whom Lawrence creatively dubbed 'Lewis'. The British Stokes mortar instructor was a young corporal named Brook. In another creative burst he was dubbed 'Stokes'. Lawrence described Yells thus:

> Lewis was long, thin and sinuous, his supple body lounging in unmilitary curves. His hard face, arched eyebrows and predatory nose set off the peculiarly Australian air of reckless willingness and capacity to do something very soon.[142]

From August 1917, Lewis was with Lawrence for around three months. He threw himself into instructing the Arabs as best he could, using a language unknown to them. He'd

---

[141] Faulkner, C.C, *Commonwealth Bank of Australia, A Brief History of its Establishment*, Chapter XII.

[142] Lawrence., p. 352.

The Arabs Have Aqaba

compensate with the use of wild hand signals, facial expressions and expletives, followed with responses that varied from thunderous assault to excited cheering, depending on the tribesmen's quickness of uptake. Lawrence praised Lewis for his affinity for life with the Arabs and his acceptance of their ways and culture.

Planning grew for an attack on the Hejaz Station at Mudowarra (about five miles from today's Jordan/Saudi Arabia border) and Lewis bounded to volunteer himself and Stokes to be part of it, even though his role was only to instruct. Lawrence immediately saw the advantage of having them there, just in case the tribesmen hesitated. Nonetheless, Lawrence felt obliged to warn Lewis and Stokes

> that their experiences might not at the moment seem altogether joyful. There were no rules; and there could be no mitigation of the marching, feeding and fighting inland. If they went they would lose their British Army privilege, to share with the Arabs (except the booty!) and suffer exactly their hap in food and discipline. If anything went wrong with me, they, not speaking Arabic, would be in a tender position. Lewis replied that he 'was looking for just this strangeness of life'.[143]

This typical Australian response was probably based on such a life being hellishly more exciting than instructing at an army school in Egypt.

They arrived at Guweira to pick up Auda abu Tay and his unruly Howeitat tribesmen. Here it was 50^0C in the palm shade where they endured fleas and ticks and were fanned by the wings of millions of flies. To dodge the daily ritual of bombs from a roving Turkish aircraft, like bats from a fruit tree, the Arabs clung off a huge rock that is still seen today with an Arab Revolt flag painted on it and which dominates the fort.

On approaching Mudowarra, they set up guns and mortars for an attack. However, the Turks spotted a group of tribesman lounging on the sand ridgeline to enjoy the sunset, causing Lawrence to abandon the idea of attacking the station with its

---
[143] Lawrence, p. 345.

superior force. So he determined to wait for a train. Fortuitously, one soon came from Hallat Ammar, the next station south, before a Turkish force could catch up with them.

The track mined, Lewis guns and Stokes mortars were set in place. The mine detonated under the second of the two engines, spilling many of the carriages into the desert sands. Lewis consistently provided accurate support fire while himself under fire from the surviving Turks defending the train. Success brought the Arabs plunder and loot with wild frenzy. Sensibility deserted them as they squealed and yelled in uncontrollable delight. Eventually they abandoned the scene with booty tossed aboard baggage camels. Lawrence and the two instructors, still with their valuable guns and mortars, were abandoned by the loot-laden tribesmen and nearly deafened by the howls of exuberant screeching.

Lewis found some baggage camels and the machine guns and mortars were loaded for escape. Weapons and ammunition that could not be loaded had to be blown up, much to Lawrence's dismay. In the escape, Lewis fitted one of the machine guns across his legs to fire as a defence against the chasing Turks. Fortunately, the exploding ammunition made such a commotion the Turks halted the chase and escape was complete.

The whole group arrived in Aqaba some days later, the tribesmen in glorious display of their booty. Lewis and Stokes were immediately thrust aboard ship for Egypt as, according to Lawrence, 'Cairo had remembered them and gone peevish because of their non-return'.

Yells was awarded the DCM, presumably at Lawrence's recommendation.

But Gaza still had to be taken. The man to do it had arrived; not a Messiah or a man of miracles as previously seen in the local history, but one with the determination and insight to develop a plan.

Part Three

A New Commander-In-Chief: General Allenby

June 1917 - October 1918

13: Gaza in Allenby's Sights

Never give an order that can't be obeyed.
– General Douglas MacArthur

Edmund Henry Hynman Allenby was 56 years old when he arrived in Cairo on 27 June 1917 and took command of the EEF. He brought with him the support of the Prime Minster Lloyd George. The historian Matthew Hughes wrote that Lloyd George recalled how, before Allenby left (London) in June 1917:

> I told him in the presence of (General) Sir William Robertson (Britain's then top soldier) that he was to ask us for such reinforcements and supplies as he found necessary, and we would do our best to provide them. 'If you do not ask it will be your fault. If you do ask and do not get what you need it will be ours'. I said the Cabinet expected Jerusalem by Christmas.[144]

Allenby, nicknamed 'the Bull', had to capture Jerusalem and win a war. His first accomplishment had been to protect his right flank by legitimising the Arab Army of Emir Feisal with promises of gold, weapons, ammunition and food after they captured Aqaba. Next, he had to take Gaza.

Allenby's arrival had been preceded with mixed feelings, so he

> disappointed neither the staff nor the troops. Within hours of taking command he made a thorough inspection of GHQ. He stomped into every room, poked into every nook and cranny, ascertained what everyone was doing. When, on the first morning at his desk, a senior staff officer brought him a formidable pile of papers dealing with detail of dress, discipline and associated matters,

---
[144] Hughes, M., p. 23.

he threw the whole lot into a corner and said, in terms described as emphatic, that he hadn't come all this way to Egypt to be bothered with details that ought to be decided by the gentleman himself, or better, someone a long way further down the seniority scale. Five days after his arrival he departed for an inspection of the front leaving behind a shaken staff.[145]

One of his early acts was to move his HQ and staff officers out of their comfortable Savoy and other Hotels in Cairo, leaving their wives and girlfriends behind. He moved them with their desks and pens to a few miles behind what would be the firing line and had them behave like military men in the field. The army applauded the move, if not the staff. It allowed Allenby and the staff to get to know, not only their own troops, but also the dispositions of the enemy and lie of the ground, somewhat essential when planning tactics and real battle conditions for a win. Times were changing.

'I could not count the times I have shaken hands with Allenby' said one light horse staff officer a few months after the leader's arrival. 'Between the canal and Gaza I never set eyes on Murray'.[146]

Endowed with a large frame, the Bull was an imposing and intimidating character. He would explode among staff or others who fell short of his expectations. Staff work improved immeasurably after his arrival and tensions between them and the Anzacs diminished, although never extinguished. Such was his energy and volatility that when he travelled between his units, to circumvent unnecessary surprises, this warning message would immediately leave his HQ for the unit to be visited: 'BL' meaning Bull loose.

In the last two months of Murray's reign, General Chetwode had reinvigorated the depressed EEF with a rotation of rest, train, patrol. Chauvel's Anzacs had fought several engagements in this campaign and won them all. New Zealand's Chaytor

---

[145] Keogh, p. 133.
[146] Gullett, p. 357.

Gaza in Allenby's Sights

had proved so capable in all tasks given him; it was thought perhaps he hadn't been challenged sufficiently. Major General Hodgson, the Brit commanding the Australian Mounted Division (at the demand of the Australian Government this had been renamed from Imperial Mounted Division) had shown sound leadership abilities and was highly regarded by the Anzacs. Major General Barrow, another Brit, who commanded the Yeomanry Mounted Division, was an old-school cavalry officer who had shown in other theatres, and since his recent arrival, that he was a mounted commander of great skill. By early June, it was known that General Murray was to be replaced, raising the morale of the EEF almost to a man. A combination of Murray leaving, rest for the men and animals and competent divisional commanders awaited Allenby.

So it seemed the EEF would have a new boss who was fair dinkum, with a job to do and a volatile manner. The first job was to take Gaza. Chetwode presented his plan, conceived by Colonel Dawnay, to Allenby: a feint at Gaza, strike at Beersheba, then advance on Gaza from its weakest points.

> Shepherded by the mounted troops and light car patrols, he [Allenby] pushed boldly out day after day on to the wide no-man's land between the two forces, carrying in his mind Chetwode's great plan, checking it in every detail, and searching in vain for something better.[147]

With these actions, he instilled a new confidence into his corps and divisional commanders, this rubbed off down the line onto the men. All ranks soon got the feeling that

> what had hitherto been a rather casual military adventure with no definite goal suddenly converted into a stern, clear campaign with nothing short of the complete destruction of the Turkish force in Palestine and the capture of Jerusalem as its immediate objective.[148]

Not every plan went as he thought they should. He noticed the horsemen were wearing shorts so he issued a directive that

---
[147] Gullett., p. 358.
[148] Gullett., p. 358.

shorts were not to be worn – perhaps concerned mosquitoes might bite and malaria would strike; or chafing on saddles; or sunburn. Nevertheless, it was hot and horsemen didn't like jodhpurs or leggings. So they simply removed their shorts and failed to replace them. Point made, bare bums became outlawed too.

The stage was set. The Bull was about to cut loose.

As he reviewed his new troops and equipment, Allenby restructured the EEF. He now had two large groups of infantry (or corps); the 20th Corps that he gave Chetwode to command, removing him from the mounted Desert Column, and the 21st Corps, commanded by Major General Bulfin, who had recently brought them from Greece.

Next he renamed the Desert Column the Desert Mounted Corps (DMC) and, to the astonishment of British officers, and probably to the soldiers too, gave its command to Chauvel, whom he promoted to lieutenant general, making him the first Australian ever promoted to this high rank.

Why was this astonishing? Well, the DMC consisted of three divisions, each with three brigades of mounted troops, and they were not all Australian. Of the nine brigades four were British, four were Australian and one was Kiwi, plus the camel brigade, with one British and two Australian battalions. It also had British and Australian light cars plus British and Indian artillery. In addition, his HQ, inherited from Chetwode, predominantly consisted of British officers; the guys who liked to be saluted.

A dominion officer in command of British troops – unthinkable! This put some class-conscious noses firmly out of joint. The DMC got more racially mixed when the British Yeomanry were later withdrawn to France and replaced by Indian cavalry.

So the upshot was Lieutenant General Sir Harry Chauvel now commanded a mix of races, religions, cultures and sensitivities amidst some resentment:

So mixed a force demanded special tact and diplomacy on Chauvel's part; he must never even appear to favour his Australians. Such criticism as there has been of his handling of this problem has come from his fellow countrymen, some of whom resented the predominantly British staff of DMC.[149]

Given the animosity between the staffers and the Australians, this is hardly surprising. Further, both Chauvel and his brigade commanders requested more Australian and New Zealand officers on HQ. Allenby developed hearing problems, the Brits stayed. Chauvel did manage to get Colonel Rupert Downes to run the medical services. This was a position he cared about deeply as he knew that health and hygiene were imperative to maintaining fitness and stamina among his overworked DMC, now a sizeable force of some 20,000 troops.

This contrasted with the poor buggers in the Turkish Army. Von Kressenstein states, 'Much of the sickness (10,000 of 40,000 in hospital) was due to insufficient nourishment as the single metre-gauge railway was unable to supply his force'.[150] The Turks, however, had it fine with water. Their defended and watered positions gave each man around four and a half gallons per day, compared with the Anzac and British troops' meagre water bottles with just one quarter of a gallon.

By now, the Turkish Army under the promoted General von Kressenstein numbered around 46,000 infantry and a few thousand horsemen of indifferent quality, plus a mix of artillery and the very capable machines of the German air force. In fact, enemy air was becoming quite a psychological problem for EEF soldiers:

> Throughout the summer, the men felt helpless in the face of enemy planes. Repeated attacks left them feeling pretty sore about the lack of any effective protection from the bombings while they remained proverbial sitting ducks.

---

[149] Hill, p. 119.

[150] Keogh, p. 125.

Desert Anzacs

> These bombs have a tremendous crashing, ripping, tearing sound and some men who were courageous against bullets and shells would tremble with fear at the approach of an enemy plane.[151]

But life was to improve for the men on the ground and those in the air. Woodfin continues:

> In September the tide of the air war changed, and a major source of fear and helplessness for the men on the ground evaporated with the effect of anti-aircraft guns [that began to arrive] ... Now they saw victories too as Allenby's new planes arrived. The first encounters with German planes were spectacular successes; as one pilot [Dickie Williams] recalled, a German plane attacked a new Bristol Fighter for the first time and, 'it turned around and bit him.[152]

So the EEF soldiers' fear dissipated but it's hard to imagine two totally different armies than the British and the Turks. In reality, the Turks were shattered:

Ill-fed, wretchedly clothed, uncertain of its supply of munitions, low in spirit, and weakened morally and physically by a continual leakage of deserters ... but the Ottoman rank and file will fight doggedly and dangerously under incredibly bad conditions.[153]

Of course, not all things happen quickly in war. Promises by prime ministers have to be met by lesser beings. So Allenby had to wait for resupply and new equipment. It would be two months before he was ready to attack. No time was lost as planning, training and rehearsals had to take place. The doldrum of incessant rides for days on end amid the heat, dust, flies and lack of sleep would weary the Anzacs. The boredom of a constant diet of beef and veggies from a tin, biscuits that dunking in hot tea failed to soften, scant leave and recreation, and only letters from home to ease the monotony.

---

[151] Woodfin, p. 76, quoting letters of two soldiers to their parents, held in the AWM.

[152] Woodfin., p. 77.

[153] Gullett, p. 372.

Resupplies did start to arrive, slowly. From early July the RFC started to receive, two at a time like animals marching beside Noah, the new version of the Bristol Fighters (Brisfits). These aircraft pushed enough German fighters from the sky to enable photo reconnaissance of Turkish movements and positions so that Allenby could fine-tune his plans.

> Control of the air was essential if Allenby's preparations were to be kept secret ... with the new Bristol Fighters, Allenby was finally able to deal with the enemy air menace and gain air superiority – a dominance he retained for the rest of the war.[154]

The Australians also received new and more powerful aircraft so that:

> Between July and October when the final battle for Gaza took place, No. 1 Sqn performed the whole of the strategic reconnaissance on the front, some of the tactical reconnaissance and much of the photography.[155]

Good maps are produced from good aerial photos and so it was here. Virtually all the maps came from No. 1 Squadron photos so the AFC was continually the advance guard for the light horsemen. 'Maps made from their photos enabled the artillery to shatter the enemy's defensive positions at Gaza'.[156] In addition, they would drop hand sketches if the enemy was close.

Another task of the airmen became an early version of PSYOPS, or psychological warfare. Leaflets dropped on towns, villages and military installations told how awful life was under the Turks and how wonderful it was under the British – written in Turkish and Arabic and endorsed by former soldiers of the Turks. The ever-loved cigarette packets were wrapped in leaflets telling similar stories. They would highlight the futility of German and Turkish plans and how they would suffer by being a part of them: better to desert now and be saved. Sharif

---

[154] Hughes, M., *Allenby and British Strategy in the Middle East 1917–1919*, p. 46.
[155] Cutlack, p. 70.
[156] Gullett, p. 71.

Hussein gave endorsements calling on non-Turkish Muslims to join his Arab Revolt. Letters were written by POWs to their families telling how much better life was as a prisoner, the food better, clothing and medical welfare better with the British compared with the Turks. These efforts encouraged mass desertions from the Turkish Army with many rewarded when joining the Arab Army against the Turks.

As well as the effort in the air, the ground soldiers captured thousand of prisoners. POWs were often interrogated for information and many provided it in the expectation that good food, clothes without holes, footwear with soles, medical attention and the avoidance of further battle would be their reward.

By now the Turkish leaders Enver and Djemal, full of cheek and confidence after Gaza I and II and remembering Gallipoli, thought they were invincible. But they failed to take account of the increasingly wretched state of their armies, thinking they could carve off some of their battalions and send them to regain Baghdad, which the British had captured a few months earlier. So now there was a huge disagreement between the young, inexperienced know-little pashas and their experienced, battle-worthy German advisers who knew from their intelligence sources and air reports that Allenby was building up his army and resources. The only way to defeat him was to attack, not by using passive defence as the pashas wanted. But the pashas had their way and relocated some of their battle-worthy troops.

Amongst all this, British intelligence somehow didn't recognise the appalling condition of the Turkish soldiers or the disputes among their leadership. Although some of the intelligence gathered was good, had they analysed the true state of the Turkish Army, a determined attack would have quickly won this campaign in 1917. But that's not what happened.

14: The Battle of Beersheba Opens Gaza

A good plan violently executed now is better than a perfect plan executed next week.
– General George Patton

The attack on Beersheba would be in two parts. The plan was relatively simple, as good plans should be:

- Chetwode's 20th Infantry Corps would attack from the southwest and take the defended parts of the town;
- Chauvel's DMC would attack from the east and take the town and water wells while ensuring there were no escapes and that no reinforcements interfered.

Speed, surprise, deception and secrecy were essential. As part of the deception, Bulfin's 21st Infantry Corps would feint an attack towards Gaza early on the day as if it were the point of the main thrust, supported four days earlier by naval and artillery fire. And a few other little acts of deception wouldn't go astray.

First, Major Richard Meinertzhagen, an intelligence officer on Allenby's HQ, rode out seeking to be attacked by a Turkish patrol. What luck: he was. Pretending to be wounded, he dropped a bloodstained case that contained a soldier's personal items, including letters and money. In reply to his 'wife's' very emotional letter about their young baby, the unfinished letter told of a feint on Beersheba with a big attack on Gaza.

Next, a light horseman performed a similar ruse, dropping a part-finished letter to his girlfriend saying what a torrid time they were having constantly reconnoitering towards Beersheba

Desert Anzacs

when really they were going to attack Gaza.[157]

As part of the secrecy, British air superiority deprived the Germans and Turks of anticipating Allenby's intentions and led to a complete miscalculation by von Kressenstein. A captured German report showed that

> in the matter of aircraft, the enemy had enormous superiority. Things got so far that no air reconnaissance was done by the 7th and 8th Turkish armies.[158]

Major defensive work soon ceased around Beersheba. Bait taken.

At this time, a couple of command changes were necessitated in the ALH. Brigadier Meredith went back to Australia for medical treatment and Brigadier Royston to South Africa for family reasons. William Grant became brigadier to command the 4th ALH Brigade and soon made a name for himself. Lachlan Wilson became brigadier to command the 3rd ALH Brigade.

The time had come. Naval and artillery fire hit Gaza as planned, four days out. On the night Bulfin's division scrambled over the sand towards Gaza. Lovely diversion.

A few days before the planned attack, Chetwode's infantry and the DMC commenced their march under those dazzling stars to avoid detection. They held up in concealed wadis during the day, amid the discomfort of heat and all that went with it, before the next night's move. Not a lot of sleep.

On the day itself, 31 October 1917, Chetwode's infantry, with the camels in support, attacked across the open rocky plains towards Beersheba's defences. Throughout the day, they made limited progress against the defending Turks and just couldn't get across the line. This highlighted the lack of wisdom of cavorting across open ground in the face of men in trenches with guns and machine guns spitting at them.

---

[157] Hill reports that these two incidents so convinced Kress that, once the attack started at Beersheba, he refused to believe the size of the British force and refused to send reinforcements. He reportedly gave an award to the patrol leader who found Meinertzhagen's letter.

[158] Hill, p. 46.

The Battle of Beersheba Opens Gaza

Chauvel had only two of his three mounted divisions, Allenby having taken one to fill the gap between Chetwode and Bulfin's divisions. He did have his Indian artillery to help neutralise Turkish guns.

There were several high, Turk-encrusted hills to the east that the DMC had to take before the final push into Beersheba. It was a long day of attacks against fierce defence. Things didn't go too well as the Turks once again showed what stubborn defenders they can be when prepared and settled in position.

At one stage, Chauvel, thinking his assaults were being held up too long, contemplated withdrawal, as the horses couldn't go on without water. 'Water your horses in Beersheba tonight' was Allenby's terse reply. Late in the day, the never-say-die New Zealanders finally captured the knob known as Tel es Sabe, critical to cover the attack on Beersheba.

Chauvel was on a hilltop with his division and brigade commanders. But now it was 4.30 in the afternoon with the sun setting at 5.00. The wells had to be taken by then. Many horsemen, after several night marches, had been at it since 9am. They were tired, hot and thirsty and tensely waiting for something, anything to happen. Three miles to town and no shelter, in the face of artillery, machine guns and rifles. The German airfield was just over the back of the town. Chauvel knew from aerial photos that the east side of the town had no barbed wire or nasty horse-trap trenches (circular pits about six feet in diameter and three feet deep, dug in lines about six inches apart that horses could not jump. They just fell in, where horse and rider were easily shot). Unable to be destroyed by artillery or air bombing, mounted troops did everything to avoid them. But there were normal soldier trenches. Time was of the essence.

The usual light horse tactics of ride, dismount, run and attack weren't going to work before 5pm. They would have to ride all the way. Who to send – British Yeomanry, rested but far away or Grant's 4th Brigade, weary but closer? 'Put Grant straight at it', became the now famous order.

Desert Anzacs

The order came, 'Regiments ... form squadron line extended ... form squadron line extended'. This was so new one trooper exclaimed, 'Shit, what's going on here?' To this his Warrant Officer casually replied, 'We're gunna charge Beersheba, mate!'[159] Grant lined up his 4th and 12th Regiments and held the 11th in reserve; 800 horsemen to charge more than 2,000 entrenched Turks. Men and boys were bursting for action. Horses shuffled:

> Two batteries of Horse Artillery galloped into action in the open to support the Light Horse and as soon as they had opened fire Grant and his regiments, squadron after squadron in line, swept into full view of the enemy, the men grasping their bayonets as if they were swords.[160]

Knowing they were Australians, the Turkish Commander, Ismet Bey, held fire, anticipating their dismount. The defenders didn't adjust the sights of their guns, waiting for the dismount. Machine guns and rifles stayed silent.

Lieutenant Colonel Murray Bourchier, CO 4th Regiment, led the two advancing regiments with bayonets held aloft, like miniature swords. The horsemen broke their horses into a trot then into a canter. And then, the unexpected 'Charge!' and the gallop was on. They hadn't dismounted.

Surprise turned to terror as the defenders madly twirled handwheels to lower their gun barrels but in vain. German aircraft did a job as the horsemen flew across the stones and flints, steel shoes showering sparks in the darkening afternoon. Flashes from the guns behind the town lit it up as they raced forward. Turkish machine-guns spewed death. Mounted but unarmed stretcher-bearers were right up there. Failing sunlight shimmered off bayonets and all the time the horsemen, bent low over their walers, screamed their bushland obscenities. Three miles became two, two became one, then only yards. The horse artillery sent round after round into the Turkish

---

[159] Wickham, H.T., 'The boy who denied his father for his country', 2002, www.diggerhistory.info/pages-heroes/boy.htm.

[160] Hill, pp. 127-8.

artillery and machine guns. By now the Turks were firing too high, even the riflemen hadn't changed their sights when they finally realised those crazy Australians weren't stopping to get off their horses as they were supposed to do. Not today, Jacko! Then the horses got the whiff of water so nothing was going to stop those walers now.

Impenetrable curtains of dust hid the battle to those on the hill. Chauvel and his commanders couldn't see – won or lost? And it was getting darker by the minute.

Two squadrons of the 12th bolted straight over the trenches into town and grabbed the wells. The following squadrons leapt off horses and were using their bayonets to great effect in hand-to-hand fighting. Many Turks threw their weapons down and their hands up. Others fought on.

Trooper Scotty Bolton, a 23 year-old from Geelong, astride his waler, Monty, was among those early into town. Monty suddenly careered wildly, nearly throwing Scotty. He later discovered a rifle shot had narrowly missed his own leg, and deflected off his water bottle but cut a swathe about 12 inches long through Monty's rump. Regaining control, on they went. Chaos reigned, with rifle and revolver shots all around. Spotting a fleeing Turk, Scotty chased, banged him on the head and grabbed his revolver. The wells had been reached and horses and riders were gulping freely. Scotty was just about to join them when he heard one, then a second explosion. Spotting wires in the sand he followed them until they went over a windowsill. Inside was a German officer at a detonation board. Scotty rushed inside putting the revolver to the German's head. One used his language and the other used his; the revolver translating instantly. Handing the German to others and disconnecting the wires, Scotty remounted only to see a Turkish gun crew making away with their gun. Not allowing this to happen on his watch, Scotty dashed after them, dislodging the driver from his carriage and forcing the others to raise their hands. At this time he needed to shoot but the click told him the revolver was empty. Fortune favours

Desert Anzacs

the bold, so, feigning leniency, he called over an officer who, with others, took charge of the gun and crew. Such creative individual initiative not only wins wars, it separates the doers from the watchers. Scotty received the DCM for conspicuous gallantry and devotion to duty.

Grant's brigade took over 1,000 prisoners with the loss of 31 killed and 32 wounded of the 800 starters – an amazing feat in the face of heavy artillery, aircraft, machine guns and rifles ensconced in trenches. Of the 31 killed, stretcher-bearer Tibby Cotter would not take another English wicket in test cricket. His grave and headstone are in today's beautifully manicured Commonwealth War Grave Cemetery in Beersheba.

During that charge, Harry Wickham charged along with his machine gun and packhorse among his mates. A Turkish bullet ripped through his femur, smashing it and throwing him from his horse, nearly to be trampled by his following horsemen. Next morning, the padre called on the injured and barely alive trooper. 'Make sure they get the headstone right' was Harry's plea, then he died. Back home, Thomas Bell received a telegram telling of his 'nephew's' death. He wrote back, 'I don't have a nephew named Harry, only a son but he couldn't be in the Army fighting, he's only 16'.[161] The headstone today in Beersheba's Cemetery is to sixteen-year-old Trooper Harry Bell. They got it right.

The message finally reached Chauvel. The town with all the wells was his. Deeply relieved, Chauvel was able to hand Allenby his prized water and the ability to now carry on the next phase to Gaza.

Ismet Bey escaped in the nick of time. His forlorn intercepted signal to his HQ said that his troops had broken because they were terrified of the Australian cavalry. Later that night a German officer claimed, 'The Australians are not soldiers, they are madmen'. Madmen? No, just well-disciplined, well-trained, well-led and full of confidence. Chauvel awarded Grant a bar to his DSO, an award that acknowledged the brigade's efforts.

---

[161] Wickham, H.T., www.diggerhistory.info/pages-heroes/boy.htm.

Chauvel had delivered again, albeit, just in time. His Anzacs had delivered again. Allenby was impressed, rushing forward to congratulate Chauvel and his horsemen.

The Battle of Beersheba is recognised as the last great mounted charge in history. Light horse and cavalry units of the Royal Australian Armoured Corps commemorate it each year on 31 October.

Figure 17: Headstone of Trooper Albert 'Tibby' Cotter, Australian fast bowler.

Figure 18: Headstones of two New Zealand soldiers, Beersheba War Cemetery

Of course, official recognition is wonderful. But what of the men who went through it, what did they think? Well, Sgt Bert Canning, age 31 at the time, who had been one of the first to enlist on 20 August 1914, described it thus:

> A dazzling success of galloping horsemen against an enemy in trenches. It was of vital significance and a shining precedent to every leader. The enemy had been beaten rather by the sheer recklessness of the charge than by the very limited fighting power of this handful of Australians.[162]

Bert thought soldiering was such a great career he stayed on and during the Second World War, was commissioned and became a major by retirement in March 1946.

---

[162] www.lighthorse.org.au/personal-histories-bert-canning.

Desert Anzacs

Meanwhile, General Bulfin's 21st Corps attacked the Gaza defences and the coastal areas to its west and north. This time, with the advantage of effective heavy naval, artillery and air support, they made good progress through the cactus hedges and barbed wire. 'On the night of 1/2 November, Australian pilots supported them by bombing enemy batteries. Williams noting the success they had aiming for the muzzle flashes'.[163] But it was not captured yet.

It wasn't all beer and skittles. Back at Beersheba, there wasn't as much water in the wells as had been expected. Although it had rained heavily in the area a week or so before this battle and many pools had water, it was limited. From these the DMC was able to partially water its scattered brigades as their operations continued. Adequate food and water were soon to become a major problem.

This light horse victory had been assisted when, in a rare moment for the wily German, General von Kressenstein had made a complete blunder in his assessment of British intentions. He'd stubbornly refused to believe the main attack was on Beersheba, rigidly sticking to Gaza. But then he got an even bigger surprise. He mistook a move by 2nd ALH Brigade on the same night Beersheba was captured, as a major advance on Jerusalem. But worse ... lo and behold, now the Arabs got into the act. Colonel Newcombe, with a meagre force of 70 or so tribesmen, had cut the Beersheba–Jerusalem road about fifteen miles northeast of Beersheba. His task had been to recruit local tribes to the Arab Revolt and to harass any uncaptured Turks hot-footing it out of Beersheba. This so jolted von Kressenstein that he now thought there was an immediate big push to Jerusalem. Unaware Newcombe's force was so small, von Kressenstein moved six battalions, including some of their best troops, away from Gaza and towards a startled Newcombe; 2,000 against his 70. Newcombe's force had only three days' of rations and was so tiny against the thousands now against them that heavy casualties and lack of food, water and ammunition

---

[163] Molkentin, p. 111.

The Battle of Beersheba Opens Gaza

forced them to surrender. But this was a job done, as Gaza now had a smaller garrison.

Von Kressenstein then moved three infantry divisions from Gaza, east towards the DMC. This left a hole in his Gaza defences that opened up when Bulfin's 21st Corps attacked in the dark at 3am on 2 November. Assisted by major artillery support and this time the successful use of six tanks, in the face of strong defence, the 21st took several key positions through the cactus, barbed wire and obstacles, but not yet the whole town.

One way to reduce deaths and excruciating pain is by providing medical care and pain relief at the earliest opportunity. In another innovative step, the Australian medical provisions for the DMC were to change ahead of British thinking and without their approval. As the movement into the desert continued it was recognised that evacuation of the wounded and sick to hospitals and advanced care required a long and rough journey. To reduce unnecessary deaths the Australian Advanced Operating Unit was established. Trains brought in necessary staff and equipment then the unit moved as close as practical to the front line to treat those they could and evacuate those who needed hospitalisation:

> This unit was commanded by Colonel C.J. Storey and had a permanent staff that did not belong to any other medical unit. This unit was employed for the first time at the capture of Beersheba. The surgical experience gained confirmed the value of the unit, consisting of several surgeons experienced in the performance of major surgical operations. It also showed that the equipment needed for the performance of such operations was neither bulky nor elaborate and that operating surgeons could cooperate advantageously with the Field Ambulances.[164]

Those who best appreciated the value of this treatment were those whose lives were saved by prompt treatment. But,

---
[164] Dolev, *Allenby's Military Medicine*, p.72.

these surgeons could only have been motivated by benefitting the wounded as, being so close to the action, the surgeons and their staff became an endangered species. Dolev reports:

> At 6am on 1st November (the day after Beersheba's capture) two enemy aircraft descending to 1,200' bombarded the area occupied by the NZMR Field Ambulance, still open for wounded and although the red cross flag was prominently displayed and there were no fighting troops within two miles, yet the aircraft, after expending their bombs, sprayed the encampment with machine gun fire; it seemed deliberate.[165]

Fingers pointed at those Germans again as their airfield had been at Beersheba until the battle began. Strangely, this behaviour was quite contrary to the chivalrous tone towards fellow airmen.

It's easy to see how the Germans won few friends among the Turks or the Allies. In a letter to his father, Australian machine-gunner Ted McCarthy wrote:

> I came face to face with a Turk, who, badly wounded, was quietly smoking a cigarette. Though I scowled at the poor devil, the Turk winked. I winked back and we both smiled. The terrible Turk is not so terrible after all. Somehow I pity the Turks – all our lads say they fight fairly – but the Germans, I detest with all the loathing that's in me.[166]

Nevertheless, the innovative changes adopted by the DMC were a credit to the cooperation of the medical staff of the Anzac Division and some time later, a cooperative British HQ.

---

[165] Dolev., p. 73.
[166] Woodfin, p. 44, referring to AWM paper PR83/161.

The Battle of Beersheba Opens Gaza

Figure 19: An Israeli family admiring the memorial to the Desert Mounted Corps, Beersheba.

15: Gaza into Palestine

Champions aren't made in the gym; it's something they have deep inside.
– Muhammed Ali

The details of the complexity of fighting aren't really essential reading. As is the nature of countryside, there were numerous small and large natural features between Gaza and Beersheba. But there was less water in Beersheba than had been anticipated. Food was a major problem so days with limited water and food had once more jaded both men and horses in the infantry and mounteds.

Compounding this, the Turks managed to just beat the DMC into Tel Kuweilfa, an important and defendable well-watered area. For five days, they held off everything Chauvel could throw at them; light horsemen, yeomanry, cameleers, the New Zealanders and infantry of 53rd Division. This was a double whammy; it blunted the Turkish forces but also fatigued the DMC. Allenby was forced to accept the opinions of Chauvel and Chetwode that rest and recovery were essential or severe counterattacks would make them vulnerable. Allenby reluctantly had to agree.

For the fortunate few, eagerly awaited mail deliveries brought some relief from the exhaustion. Stories from home and loved ones meant a great deal to the men who had little time off to ease their troubled minds, and their bodies that ached from the neck, shoulders and arms, to the back and hips and legs. Many took the opportunity to write their own letters and postcards, diaries and stories. But it wasn't all rest. Regrettably, burial details were a last chance to talk to mates who were never to walk or ride their course, and with whom they would never again share another joke or cigarette. The care of their walers was a high priority, rated well above their own. After a few days rest and widespread attacks of diarrhoea from gulping local water when nothing else was available, on went the war.

In the heat of battle, mistakes can occur. The Red Cross and ambulances were sacred to both sides. Turks were generally a respectful enemy, which is why this incident was met with outrage:

> As a rule the Turks scrupulously observed Red Cross rules; but on this day all enemy arms fired very deliberately upon three ambulances and carts, which had been sent up over the exposed ground for the wounded. The carts were clearly marked with the Red Cross, the visibility was good, and the Turks were shooting at close range ... most of the drivers and wounded were hit and horses killed ... A Turkish doctor, taken prisoner soon afterwards ... explained that the ambulances had been fired upon by the urging of a German officer, who argued that the carts were probably carrying ammunition up to the Australians.[167]

Such contemptible behaviour indicated the barbarous attitudes of some Germans that would be compounded during the Second World War. German airmen, in an earlier encounter, showed their chivalrous, almost sporting nature towards our airmen, if not our hospitals:

> On July 8, a patrol of three went out from No. 1 Squadron – Captain Brookes and Lieutenant Vautin as escorts to Lieutenants Taylor and Lukis as reconnaissance. Near Gaza two German scouts attacked the escorts ... the wings of Brookes plane were seen to fold and his tail fell off, the machine going down like a stone ... Taylor met the challenge with two well-directed bursts of his machine gun ... the German turned and attacked Vautin. Taylor and Lukis reached home but Vautin was forced down and taken prisoner. The amiable Felmy sent a letter two days later to say Brookes had been killed and was buried

---

[167] Gullett, p. 417.

with military honors. Felmy wrote that Vautin was quite well, and hoped we could send him some kit.[168]

Felmy, the German air commander, bestowed other courtesies on Vautin, who wrote notes to his mates that were later dropped by German pilots on the AFC airfield:

> Following receipt of these interesting messages where war was carried on with such old-fashioned chivalry, Captain Murray-Jones flew over the lines with Vautin's clothes, small kit and home letters. Felmy and other German airmen were waiting for him on their aerodrome. Jones descended as low as fifty feet and dropped the parcel among them, then circled the ground, returned the enemy's hand waving, and flew home. No shots were fired at him.[169]

Back to the less chivalrous fight. British infantry of 20th Corps under Chetwode, together with the cameleers, light horsemen and British cavalry, captured several objectives north and west of Beersheba. Kuweilfa held out. British and Australian airmen took photos, rained bombs and strafed. Such was the efficiency of the aerial photos now that these valuable pieces of intelligence were available to the ground commanders within hours. Nevertheless, the lack of water and food, the casualties suffered, the need to extract the wounded and dead from the field, the resupply of ammunition and insufficient rest or sleep wore down every man and every animal. The medics worked feverishly, the vets were tireless, transport drivers and camel handlers were ceaseless.

It was the same for the Turks. Artillery, machine guns, riflemen all fought on. But the tide had turned. The battered Turks, heavily outnumbered from the beginning but resolute in their defence, started to see the wisdom of discretion. Many a Turk headed for the back door: they simply deserted. By 6 November, many Turks were on the run. The RFC ground crews rushed to load planes with bombs and machine gun rounds.

---

[168] Cutlack, p. 72.

[169] Cutlack, p. 72.

The German airmen were without communications to the front and had no idea where the EEF was. One report suggested the cavalry was two kilometres away and pressing hard. The Germans, also favouring discretion, loaded what they could onto trains and burnt the rest. Wrong report, but next day a patrol of No. 1 Squadron spotted the departing Germans and the 40th Wing sent up 22 bombers with Brisfit escorts – carnage. Von Kressenstein recorded that 'This raid did more to break the heart of the 8th Army and to diminish its fighting strength than all the hard fighting that had gone before it'.[170]

Rumours of British cavalry at the outskirts caused massive Turkish desertions and

> many formations began to retreat without orders and broke into flight. A great number of officers and men could not be stopped until they reached Jerusalem (fifty miles away) or Damascus (300 miles away). A fortnight later, German military police arrested 6,000 deserters in Jerusalem.[171]

On 7 November, Gaza fell. The Turks withdrew hastily and beat a retreat towards Jerusalem. Bulfin had no mounted troops so a chase wasn't possible. The Turkish troops that had held Kuweilfa withdrew. All along the line Turks and Germans bolted. The DMC, short of water and food and lacking rest or sleep just couldn't chase quickly enough to cut them off and capture the decimated army. And as the Turks withdrew, they'd damage or spoil as many water sources as they could. Captain Robert Elwood recalls a horror experience:

> On this occasion we approached this hod just the horses and everybody anxiously looking for water so we didn't drink our two water bottles and we smelt the stench from sheep and goat dung and when we got to the wells all the sheep and goat dung around them for years had been raked up and put into the wells and the water smelt

---

[170] Molkentin, p. 113, quotes these reports dated 21 November 1917 from Bavarian Flying Sqn 304 and a report from major von Papen to von Falkenhayn.

[171] Molkentin, p. 113

> ... and I don't like saying it ... it tasted like nothing and I saw the horses then vomiting the water up and we had to drink it and the horses had to drink it, I'll never forget that.[172]

Yet, every now and then something pleasant could occur to ease the pain for a while. Ellwood also recalls:

> Whenever we were on the desert, almost, wherever we were, and we pulled up for lunch or a rest period, out of the empty desert came boys with oranges for sale, native boys. How they got there and where they were before, we don't know, but they were always there.[173]

The fresh fruit was heavenly nectar, a gift from God, in a diet in which everything was tinned. The juice would wash away layers of encrusted dust from lips, for a moment of comfort.

When the troops did enter Gaza they were once more astounded and horrified at the way the Turkish soldiers could live among debris, unsanitary and unhygienic conditions, surrounded by waste and rotting animal carcasses, filth and evil stenches, yet survive and fight stoutly under conditions impossible for western soldiers. Medics and engineers had to move in fast to sanitise the place and make it fit for allied occupation, while the Allies did their best to move out fast. The mud hovels had been severely damaged by the bombing of the artillery, navy and air force. But, understandably, much of the structural damage to the buildings had been caused when the Turks themselves pulled off timber, masonry and anything else that could strengthen their trenches and obstacles. Moreover, British gunfire had pelted an old mosque, originally a twelfth century church, partially destroyed following an act of age-old behaviour by unconscionable defenders, performed even today by terrorists and rebels. The Germans had used it as a munitions store believing it would not be fired upon. Once its use was known beyond doubt, fired on it was.

The pursuit continued at the best pace that could be

---

[172] Ellwood, R., *Foul Water*, p. 23 (removed from site).

[173] Ellwood, R., *Food in the Desert*, p. 22 (removed from site).

mustered. Chaytor and Hodgson's divisions went on even though most of their horses were spent, yet still gave more. Ahead of the two divisions were around 12,000 to 15,000 very confused and exhausted Turks. Had the divisions had swords rather than heavy rifles and short bayonets, 'they might have demoralised and captured the whole Turkish force'.[174] To the rescue of the Turks came the German machine-gunners, 'watchful, fearless and straight-shooting'. They prevented the encirclement and surrender of these Turks, so prolonging the campaign by another year.

Easy to see how young men could be attracted to flying. The pilots had access to endless clean water, a comfortable bed each night, a shower every day, a beer in the mess in the evening. They worked at a cool altitude, used horsepower that didn't tire as the mechanics fixed their engines almost instantly, and they had an ever-present willingness to jump aboard their plane and go bomb a Turk. While the mounted troops were exhausted, the airmen were not. That's not to say it was all easy, especially when you don't have air superiority and the abrupt end of a crash and death or capture can ruin your day. However, from Gaza,

> in eight days, the Turkish armies had retreated 90 kilometers, harassed all the way from the air. Most of the bombs fell on choke points in the enemy railway system, such as Junction Station, to slow their retreat. The British and Anzac mounted troops pursued them, but supply problems impeded their progress and prevented the decisive breakthrough and encirclement that Allenby envisaged.[175]

Allenby won a great victory through his combined use of foot soldiers, mounted troops, naval and artillery fire, a few tanks and armoured cars with his sound use of air power and the huge effort from his medics, vets, drivers, signallers and engineers. So too the dash of his commanders and his use of

---

[174] Gullett, p. 445.
[175] Molkentin, p. 113.

deception and surprise, which caught the enemy off guard so that they committed their main force in the wrong direction and committed all their reserves early.

Quickly becoming apparent to Allenby was the professionalism and dependability of Chauvel and the Anzac troops of the DMC. He confirmed his faith in appointing Chauvel to lead the mounted troops of all nations.

But his victory was not yet complete.

Beersheba had been captured on 31 October, Gaza on 7 November. On 11 November, Allenby received a cable from the War Office. Things were not going well in Europe, he was told:

> While no opportunity should be lost for weakening the enemy's defensive power, by the summer the British forces in the East may have to be reduced to the minimum required for defensive purposes ... in the meantime, exploit your success to the utmost.[176]

This was political-speak for you will lose your best troops come spring so do the best you can while you can. November was winter, so Allenby had four to six months to achieve his quest to annihilate the Turkish armies and 'exploit his utmost' in Palestine, and six weeks to win the Christmas present of Jerusalem.

On 16 November, Allenby called in his three corps commanders to a conference. He gave them the news about Europe then explained:

> In seventeen days [since Beersheba] the Gaza-Beersheba barrier has been shattered, some 10,000 prisoners, many guns, and much material captured, and the enemy driven in disorder for nearly 60 miles.[177]

But Allenby wanted more. The chase had been hampered by lack of water for the horses in the critical week after Beersheba while his men fought by day and moved by night, chasing and capturing stragglers, then having to care for their

---

[176] Gullett, p. 486.

[177] Gullett, p. 486.

horses and themselves. The men were shattered. Fortunately, the Australian transport and stores services, directed by Colonel Stansfield, regularly brought food and ammunition to the very backdoor of the mounteds, giving their weary bodies a new lease on life, enabling them to press on with the demands of battle. At one stage, some horsemen arrived in Von Kressenstein's abandoned HQ and, to their delight, their quivering nostrils found a bakery right next door with fresh, warm bread just out of the oven. A fortunate few feasted.

On the same day as Allenby's conference, Chauvel established DMC HQ in a Jewish village called Khirbet Deiran. Here they enjoyed the hospitality of the friendly inhabitants for the next five months. As it happened:

> In the same settlement, surrounded by its orange groves, vineyards and gardens, were the parents of a trooper from the 10th ALH Regiment. Chauvel immediately sent for the soldier, who was soon being feasted by forty-five of his family and friends.[178]

It's a special commander who realises the joy of finding family connections in a far-away place in the middle of a war then makes sure that a reunion takes place to create a moment of family love and sharing. This, to a large extent is what we go to war for: to create a peace – funny thought, really.

Soon thereafter, another reunion took place when the New Zealanders entered another Jewish village, Wadi Hanein. A woman asked a New Zealand officer if he knew her son. By chance there was within earshot a soldier with a letter in his pocket for her from her son. A quick message went to the Auckland Regiment and the son came galloping over into mum's arms and an emotional welcoming. Imagine what joy and heart-warming impact two such meetings could have had on soldiers and families in such circumstances.

Chauvel was the type of leader who led from the front and was in this way an exception to some of the other commanders. The welfare of his soldiers was always at the top of his mind.

---

[178] Hill, p. 134.

Figure 20: An original armoured car, restored and displayed in Royal Auto Museum, Amman, Jordan (with permission).

At Beersheba, he'd been on a hilltop to view proceedings and direct the battle, within enemy artillery fire and in a place where a German aircraft dropped a bomb so close that dust landed on his table, although no other damage was done. He was constantly in between his units and talking with his officers and soldiers. 'I live mostly in a Rolls Royce these days', he wrote to his wife. But this day he was on a horse:

> The fourth Christmas of the war came in, not with traditional snows, but with rain and gales. So bad was the weather Christmas Day, that Chauvel rode around the camps to visit his troops and share their hardships. With him rode Brigadier Trew [a British officer, his administrative head of staff]. As they approached two bridges over a wadi, Chauvel saw a horse in serious difficulties on one of them, a railway bridge; the animal was down, with its legs between the sleepers. Leaving their horses with his orderly Chauvel and Trew picked their way along the track, the rain pouring and the flood roaring below. With the help of Trew and a soldier, Chauvel secured the horse's legs, rolled it onto its

back and then over again so that it fell into the water and quickly made its way back to the bank. Trew was much impressed; later that day on another ride, he was surprised to find a second horse in the same unhappy fix. So Trew, the Royal Marine, went to the rescue and did all things needful, except that he omitted to jump clear as the horse rolled; its legs caught him and he accompanied it into the water.[179]

As 1917 closed, No. 1 Squadron AFC wasn't to be exempt from depletion. Not only did No. 1 have inferior aircraft to the Germans and their British mates, but now AIF HQ wanted their experienced pilots and observers to bolster the three new squadrons being sent to France. They also wanted No. 1 Squadron's experienced mechanics. Furthermore, the British pilots and observers who had been with No. 1 Squadron, as trainers and to make up the numbers, were also withdrawn to bolster the new RFC squadrons being formed in France.

So new pilots, observers and mechanics had to be found and out went the 'positions vacant' notices:

Since the beginning the squadron had enlisted volunteers from the light horse, and horsemen were the finest possible material for the work required. As the British officers left, their places were filled by Australians, the majority from the Light Horse.[180]

And the squadron commander, knowing replacement mechanics could be quickly trained, gave his older mechanics the opportunity to fly as a reward for excellent service.

In a further boost to Anzac morale, Allenby brought with him a new approach to honours and awards; a more even distribution, based on merit over place of birth. Chauvel again wrote to his wife, 'Lots of decorations have been earned by my people since Beersheba under the new regime. I myself can give out Military Medals now'.[181]

---

[179] Hill, p. 133.
[180] Cutlack, p. 69.
[181] Hill, p. 134.

Chauvel's leadership and tactical successes saw him recognised in the 1918 New Year's Honours List with the award of a Knight Commander (KCB), an upgrade to his Companion to the Order of the Bath (CB).

Allenby's target was now Jerusalem, a city that sits among steep hills and rocky plateaux. These are impossible for horses, camels and wheeled wagons towing artillery and stores to negotiate and tough going for infantry with pack mules and donkeys. Captured in 1187 by Saladin and ruled by Muslims for 730 years, the last 400 years by the Ottomans, Jerusalem was to be a morale-boosting Christmas present that PM Lloyd George wanted for his devastated British public. Allenby's EEF had to do it.

The British infantry were tasked to capture Jerusalem in what could be a very costly stoush to those young soldiers. It was expected that casualties would be high and the logistics a nightmare. The mountains, hills and valleys around Jerusalem were like violent waves on a troubled ocean. Winter rains, cold weather, mudslides and slippery rocky grounds greeted the English farmers and shopkeepers. The Turks were once more well established in strong defensive positions with artillery and trench works presenting enormous hurdles for an attacking force. The British mood was apprehensive. With great difficulty, their own artillery was dragged into position. Holy sites were likely to suffer, as would the large civilian population.

By 8 December, the British were ready to strike the following morning. By now the RFC and AFC were able to operate with greater freedom. But the storms and rain on this day meant there was no offensive air cover for either side. Pre-dawn awakenings found the cold and sodden troops ready for fierce combat. The artillery had been pounding the hills and earth for hours while Turkish artillery was strangely quiet.

Early air reconnaissance, the only aircraft that could fly, went up. The word came back – the Turks had abandoned Jerusalem. They were gone. Apprehension turned to bewilderment then jubilation flooded the weary, sodden and scared soldiers. They entered Jerusalem without a rifle shot, without damaging the holy sites, without a single casualty to the civil population. A great sigh of relief almost blew away the storm clouds.

General Shea, commanding the British infantry division, entered the city with a guard of his wet and muddy Londoners, met the Mayor and graciously accepted the surrender of Jerusalem:

> General Shea was greeted by shouting and almost hysterical crowds, more various in race, religion and speech than are to be found in any other city in the world ... Filling every window and densely packed on the flat roofed houses, they welcomed the little party of Englishmen in a babel of many tongues.[182]

The residents were more elated than the soldiers. Seven hundred and thirty years of Muslim rule and 400 years of Ottoman rule were over.

The first Australians entered Jerusalem that afternoon. Major C.G. Dunkley rode at the head of his Squadron of the 10th ALH Regiment, Western Australians with their emu plumes flapping in the breeze and to the bounce of their walers:

> They were rushed by the populace, who marvelled at the size of their big, long tailed horses. Accustomed to the slender Arab ponies, the people of Palestine and Syria were always impressed by the height and power of the splendid animals.[183]

The British Prime Minister and War Office were jubilant. The news was published throughout Great Britain and flew down the airwaves across the world. Christians, Jews and Muslims were all jubilant, as they could once again access their sacred places.

In contrast to the gaudy entry by decorated carriage and horses in 1898 of Kaiser Wilhelm II, when the city walls and Jaffa Gate had to be removed to enable his entourage to fit through, Allenby walked into Jerusalem on the morning of 10 December

---
[182] Gullett, p. 517.
[183] Gullett, p. 519.

to meet the assembled population. Soldiers of all nations of the EEF lined the streets as a guard of honour to the city. The people of Jerusalem acknowledged the dignity of Allenby and rejoiced. Allenby delivered and posted his Proclamation

> in English, French, Arabic, Hebrew, Russian and Italian from the entrance to the Citadel, below the Tower of David built by Herod in 25 BC. These sites were to be guarded and preserved for the free use of worshippers.[184]

The proclamation gave free access to the city to people of all religions, nations and cultures – it would be an open city under temporary military governorship.

Again it became apparent, that Turkish domination of Jerusalem had left the city in a state of indescribable filth, disease, poverty and with broken-down utilities and infrastructure. Sanitation, garbage, healthcare, water treatment and spoiled food had to be treated for both civil health and the soldiers. The need for rest and recovery of the soldiers who'd been given leave was spoiled by the foul concoctions offered as alcohol, the frightening amount of venereal disease, thieving by inhabitants, filthy and unhealthy hotel rooms, unscrupulous traders and the dilapidation of former beautiful holy sites. Captain Hector Dinning reports:

> The filth of the city you could overlook; but you cannot escape the filthy utility by which the Greeks, Latins, Armenians and Mohammedans have transformed the ancient Holy Sepulchre and the Temple Area and the Mount of Olives and Gethsemane into a source of gain. The Holy City is infested with glib showmen who have no feeling for the past ... there is nothing sacred to them.[185]

Although the light horsemen and cameleers played no part in the capture, they took their leave in this ancient city and were appalled. Dinning once more:

---

[184] Dinning, p. 55.

[185] Dinning, p. 56.

> The Arab women, squat on their haunches, will sell you figs, melons, pomegranates and grapes – you take them, in the midst of the flies and horsedung, from the foul hands of these hags.[186]

Nevertheless, many of the Anzacs were keen to see the biblical places they had heard about since childhood. Many of the troops were disappointed by the condition of the biblical city, although, when their eyes fell upon the street stones that Jesus and Roman legions had trodden 2,000 years earlier, or the Church of The Holy Sepulchre, where Jesus was crucified and resurrected, they came to a spellbound halt. When they saw the city walls surrounding the ancient buildings, the remnants of the sacred temples, the golden dome, the stone in the Al-Aqsa Mosque from which Mohammed is said to have leapt into heaven, they may have realised, just as people do today, that, just maybe, the old stories were true. Many young soldiers went home with trinkets as souvenirs of their visit, indelible images and the recognition that a higher power exists. Such is the aura cast by Jerusalem.

But the Turks weren't done. They counterattacked several times. On each occasion they were beaten back. The superior British air power again proved decisive in thwarting ground troops. Airmen and soldiers rejoiced. At long last the advantage was theirs and fewer German air attacks tested their courage or troubled their minds.

Now the Turks had been driven from Gaza and expelled from Jerusalem. Their prestige was falling and these events were described as a political calamity in their own writings. By the end of 1917 the Turkish Army had been fractured into two parts. One force was to be found on the plains towards the Mediterranean and was in bad shape. Another was east of the Jordan River and Dead Sea, in the rugged mountains and facing the horrors of heat, flies, malaria and dysentery in the Jordan Valley. But this was winter; rain, cold, mud. The Jordan Valley would wait. And the Anzacs endured the horrendous

---

[186] Dinning, p. 52.

Desert Anzacs

cold of the hills before enduring the horrendous heat of the valley.

Allenby met his PM's challenge. Jerusalem was delivered before Christmas. Meanwhile, Chauvel turned his attention to the issues of commanding all the Anzacs under his command, from soldier and animal hospitals in Egypt to the weary troops in Jerusalem. He was greeted with many good issues. But there were some bad issues that only youthful and excitable men released from the strain of fighting for their lives in some of the worst conditions of weather and terrain the planet could throw up. The good issues included the rest, leave, rehab, resupply, wounds and sores treated, letters read and written, horse races and gymkhanas against the British Yeomanry and just goofing off. The bad issues included drunkenness, venereal disease, biffs with the lads from other nations, still not saluting Pommy officers, kicking a few unscrupulous traders and generally mucking up. Some got leave to the Mediterranean to go to the beaches and swim in the sea. Some got to Egypt where class distinctions separated officers from soldiers and many decent hotels, cafés and bars were out of bounds to mere soldiers at the direction of the unlovable British staff. Imagine the fun the military police had ejecting Anzac soldiers from 'officer only' establishments. To provide alternative, non-threatening recreation, two non-military Australian women, one a 26 year-old and the other near 60, won the gratitude of many a soldier and wandering visitor. Verania McPhillamy (the younger and commonly called Rania or Rainbow) and Alice Chisolm

> set up the Empire Soldiers Club in Port Said, and later, Kantara and Jerusalem. The clubs offered rest and recreation facilities for men on leave from the front, regardless of rank or class.[187]

Any nationality was welcome and in 1917 the daily average number of visitors was 2,500 during the 24-hour service. As the Beersheba battle in October 1917 approached and troops were given rest in preparation, the daily average soared to 4,500 delighted soldiers.

---

[187] Horsfield, J., *Rainbow: The Story of Rania Macphillamy*. pp. 80-1.

These clubs offered resting soldiers a substantial meal, a shave, a shower, and a place where they could buy writing and reading materials, food items, smokes and sweets. With some ingenuity and amazing donations of things 'found' the staff were creative; one staff member could 'pedal a fixed bicycle ingeniously connected to a drum-like contraption for churning ice and cream and fruit into real ice cream'[188], in the desert. With a gift for making contacts for obtaining supplies, the adored owners and their crew regularly served fresh vegetables and meat, much to the delight of their guests.

Further, as Christmas 1917 arrived, Chauvel and some of the Anzacs were royally feted by their Jewish neighbours. In one village 'they decorated a hall for use as a church and made gifts of wine for the Communion'[189] – it's likely more than a few horsemen became religious that Christmas. And as the New Year came and their rest continued, the Anzacs looked over the groves of oranges, almonds and olives, and over the fields of sparkling wild flowers with iris, daisy and cyclamen, anemone and narcissus as they bloomed into spectacular colour to ease the vivid memories of the bare sand and desert of the Sinai, and the rigours of fighting before and after Beersheba. They had earned their moments of peace and tranquility.

Pretty soon that was over, too.

The Turks seemed to be on the back foot but no one thought the war would be over any time soon.

Next, the Russian Revolution of November 1917 saw the Russians exit the war by December. Released from the Russian front, Germans in large numbers moved to France. That simply added to British fears, already reeling from the carnage of Ypres earlier in 1917, the more recent carnage at Flanders, and still reeling from the tactics of their incompetent generals, still unable to win. So they took more experienced and trained infantry and artillery from the east to France to counter this new German onslaught.

---

[188] Horsefield, p. 87.

[189] Hill, p. 141.

Desert Anzacs

Allenby now had to contend the weather plus the drain of his valuable fighters. They wouldn't touch the Anzacs though: they stayed. In the place of the British troops arrived untrained Indian cavalry and infantry that he would train, equip and motivate.

In the political arena, Prime Minister Lloyd George sacked General Robertson with his Franco-centric focus and anti-eastern attitude. He replaced him as CIGS with the easterner General Sir Henry Wilson, a top British soldier and representative at the newly formed Supreme War Council in Versailles, which included the French and Americans, to strategise the rest of the war:

> The predictable outcome was that Allenby's orders in early March were to press his advance. Lloyd George wanted to see this advance completed quickly. With his eye firmly on the post-war, the premier had judged that Syria would be invaluable in peace negotiations, and the capture of northern Palestine was the first step towards that goal.[190]

So,

> the first intention of the British High Command was to prosecute the campaign in Palestine with all vigor and to continue the attack towards Damascus. In view of the threat of the Germans against the Allied front in France, however, it was decided that no further reinforcements could be sent to Allenby for some time; and Allenby on his part was determined not to continue his offensive until he was assured of adequate support.[191]

Palestine came to a temporary standstill. However, it was winter and no further advance of note could be made in the rain, snow, cold and mud. The ground had to dry out so hooves and wheels didn't sink. Months of preparation, resupply and renewed training became standard. And, of course, the Turks, too, could recover, resupply and prepare defences as best their

---

[190] Woodfin, p. 111.
[191] Cutlack, p. 92.

worn out logistic system would allow.

Concurrently, by coincidence rather than battle coordination, the NAA progressed towards the southern region near the Dead Sea and the Hejaz Railway, captured several Turkish positions thanks to their British and French advisers with explosives, machine guns, aircraft and artillery. On a couple of occasions, the Arabs themselves won victories and took significant prisoners and captured weapons, to the chagrin of the Turks still in their southern bunkers in Ma'an and Medina. The Arabs could almost smell and touch Damascus: their political target too.

Allenby knew he couldn't move north towards Damascus for months – the exigencies of the weather and the need to train his new Indian arrivals slowed him down. But his star sign was Taurus, and the Bull was not a patient man. He wanted to move east, capture the important Hejaz Railway station in Amman (now the capital of Jordan) and finally isolate and force the submission of those Turkish garrisons at Ma'an and Medina. He could then move north along the railway to Deraa to prevent any uncaptured Turks supporting Damascus. Perhaps he could entice the Arabs into a worthwhile action on the way into the Jordan Valley and Jericho, offering them a path to Damascus.

The plan was to go through Jericho, the Jordan Valley, then east across the Jordan River and into the mountainous hills towards Es Salt, Amman and the Hejaz Railway. The Jordan River today is a slimy, mud trickle of the water redirected by Turkey, Syria and Israel long before reaching Jordan. But a hundred years ago, it was a raging torrent, especially after winter rains.

The Jordan Valley runs north to south. Although a wonderland of biblical stories, it's one of the most unpleasant places on earth. It descends from Jerusalem through tangles of ravines, gullies and cliffs on the west, descending from around 2,000 feet above to 1,300 feet below sea level, into a summer furnace that Trooper Robert Bygott could equally have described as being so hot it could melt the sins out of

Desert Anzacs

Satan. Further east, the Jordan River connects the Sea of Galilee with the Dead Sea. Here it splits the valley so that to the east are more hills and mountains with their ravines and cliffs so steep and difficult to travel that in many places it's not possible for horses or camels to pass, even on a dry day. And when it's a rainy day it's so slippery that a horse or camel has trouble getting up or down, even one at a time, and in many places it becomes impassable.

Atop the eastern cliffs, not far from today's Madaba with its renowned mosaics and ancient churches, is the Mt Nebo memorial of Moses' first view of the Holy Land. Moses had led his people to the land of milk and honey, gazed west across the Jordan Valley and pronounced goodness-knows-what to what you'd guess was a startled gathering of followers. Mind you, some thousands of years ago there had been pastures, trees, gardens and grazing wildlife; but for the Anzacs, those natural riches were long gone leaving a desolate landscape. So the Anzacs had the opposite opinion: they were about to leave the milk and honey and head to Moses' lookout.

By mid-February, Allenby was ready to strike. First up was Jericho. Nearby, General Chetwode commanded his infantry force. He also had the Anzac Mounted Division commanded by Major General Chaytor with Brigadier Cox's ALH Brigade and the NZMR Brigade. Chauvel and the rest of the DMC looked on, out of the action. The horsemen were tasked to cut the road to prevent Turkish support getting in and anyone getting out. And as they approached this ancient town rich with biblical history,

> they rode down through the old Wilderness of Judah with thoughts of John the Baptist and the Forty Days of the Fasting, and no tourists were ever more eager for a first glimpse of the Dead Sea and Jericho and the Jordan.[192]

Imagine how strange they may have felt as they prepared for battle. They had recollections of the Sunday school stories

---

[192] Gullett, p. 538.

of the biblical lands that had seemed to be no more than figments of a preacher's repertoire and Old Testament stories. And now, here they were, where fighting started thousands of years before and was happening still.

As it happened, the Turkish Commander figured his fighting should end. In the dark, he split his force into two groups; one headed north and away while the other headed east across the Jordan. In the first light of morning, Lieutenant W.C. Kelly of 3rd ALH Regiment led his troopers into Jericho uncontested, took some prisoners, discovered the water supply was intact and reported that the Turks were gone. And, as in previous captured towns, the picture was repeated

> there was nothing in Jericho to excite the admiration of the light horsemen. The squalor of the mud and straw hovels was relieved only by the minaret of a small mosque, and by the wretched Jordan Hotel, which in days of peace had housed tourists unfortunate enough to travel without their own camping equipment. To the habitual filth of the natives was added the filth of the Turkish soldiery.[193]

Such scenes were now commonplace and represented an alien lifestyle. The Anzacs and British infantry were greeted with a desolate wasteland; gone were the renowned gardens and pastures of Roman occupation, replaced by the repetition of the stench and images of exposed garbage, broken sanitation, unburied animal carcasses and little sign of community values. Yet 'like many other places in the Holy Land, it deeply stirred the emotions of every soldier in Allenby's army'.[194]

Unlike the Anzacs, the Turks were happy to leave their sick and wounded behind, probably comfortable that EEF medical treatment was a hell of a lot better than theirs.

> In Jericho were found a few Turkish wounded and ten [with] typhus fever. All were in the neglected state of starvation and dirt usually met within an abandoned

---

[193] Gullett, p. 540.

[194] Gullett, p. 541.

Turkish hospital. The typhus cases were in the basement of the Jordan Hotel, left to fend for themselves, with their food passed to them through the windows.[195]

It's no surprise that Turkish soldiers deserted in large numbers or surrendered quickly when they were overcome.

With Jericho captured, the Anzac horsemen were relieved by the British infantry and returned to the DMC at Bethlehem. Perhaps this move conjured recollections of stories of the Holy Land, where maybe a few even sought a manger. Diaries, notebooks, letters, postcards, photos and memories guaranteed that their family and friends back home shared glimpses of what impressionable soldiers could interpret from a land troubled over the millennia by alternating peace and hostilities. Both locals and soldiers may have asked, what's normal, peace or war? That question is still unanswered.

Allenby continued the extension of the railway, improved the road systems and brought forward airfields. This ensured the continuation of his supply lines, vital to a force on the move or awaiting the next move. The medical services brought up their CCSs, advanced surgical units and stationary hospitals serving the sick and wounded. The veterinary services brought forward their MVS for those beloved horses and nearly-loved camels. Alice Chisholm and Rania McPhillamy were given permission to set up a new canteen of their Empire Soldiers Club in Jerusalem for the soldiers out of the front line.

---

[195] Downes, p. 680.

16: 'Raids' Across the Jordan

Some days are diamonds, some days are stone.
– John Denver

It was 1918. By March the rains ceased, giving the drenched troops temporary respite. With sunshine the reds, yellows and blues of the wildflowers and the greens of the grasses poked out to dazzle their eyes, worthy of writing home about. Soldiers' clothing and equipment dried out. The livery of the horses and camels dried out. Some relief was at hand. Smiles returned to sodden faces. The Indians, indeed all EEF troops, maintained their training and re-equipped in preparation for what was to come.

Allenby's intentions were to secure his right flank, advance north along the Hejaz Railway to cut Turkish supplies to their troops in the south and prevent these southern troops from interfering. That railway was the only means of Turkish transport. Their road transport was non-existent. He would also link up with the Arab Army permitting it to make a worthwhile contribution. First, he had to capture Es Salt then Amman (the ancient Philadelphia, where a well-preserved Roman theatre can be seen today, which is still used for classical and boisterous rock 'n roll concerts). Further, the success of doing so, would go a long way to enhance the confidence of the ever-doubtful Bedouin tribes. It was an ever-present threat that these tribes could be swayed towards the Turks.

Allenby also had a higher goal imposed on him by the new attitude of Lloyd George and the new CIGS General Wilson: a final assault through Palestine and the total destruction of the Turkish armies. But, just as his planning was getting up a head of steam, along came the German offensive in France, so, once again, his best troops were poached by the war in France. His own major assault was again delayed.

Now the Turks and Germans were busy too. General von Kressenstein had been blamed for the loss of Beersheba and Gaza, so he was sacked just as Dobell and Murray had been. General Liman von Sanders was brought from his HQ in Turkey to Amman to command all troops, together with a build-up of German troops and armaments. Since he had been effective at wining Gallipoli, he was expected to do so again.

With remarkably uncharacteristic poor planning and staff work, Allenby came up with what must have been the worst plans of his career. He tasked General Shea and the British infantry to cross the Jordan and head for the railway station at Amman. Shea also had Chaytor's Anzac Mounted Division and the camel brigade, supported by supply caravans of the CTC. At a critical time the rains returned. On 21 March, 'Shea's Mob,' as it became known, set off in a howling wind and sleet across the flooded Jordan to ascend the slippery-slide of the 4,000-foot mountains to the town of Es Salt. Salt had a predominantly Christian Arab population known to be friendly. Then on to Amman to take the railway, destroy critical bridges to the south and close off rail movements north and south.

Days of delays in starting gave advance notice to the Turks who repositioned their troops and the assault was a total disaster. In the appalling weather, the AFC and RFC aircraft could barely fly and were unable to give accurate information of the Turks' dispositions, couldn't direct artillery fire and were not able to forewarn troops on the ground of enemy movements. The guns of the artillery had to be manhandled up that 4,000-foot rise from the valley, through the precipitous cliffs and gullies because animals couldn't pull the heavy guns. In making their advance,

> camels slid and slithered all over the place. Pack animals laden with explosives gave drivers hair-raising possibilities every minute. Finally the cameleers

Figure 21: Hejaz Railway station at Amman, captured by the Anzac Mounted Division, September 1918.

> dismounted and led or pulled their mounts all through the night ... one patch of mountain track was so precarious it took twelve hours to cover 200 yards.[196]

In the dull light of a day and the dark of night with rain pouring down, drenched and hungry, the men endured cascades of water running down their backs and into their boots. They were splattered with sticky mud in much the same way as those in the trenches of France. To raise themselves above the mud, they slept on beds of rocks using stones for pillows. They were shot at in the freezing cold and fell over in the mud, carrying their heavy rifles and equipment while dragging their beasts along. In fear, many wondered if they would ever see another Australian skyline. Nevertheless, these

---

[196] Langley and Langley, p. 118.

Desert Anzacs

youngsters stuck their task with the courage and determination that only a man who has known the worst can bring himself to muster and keep going.

Many EEF troops were killed. Many more were wounded. Others were simply exhausted. And it was impossible to use wheeled ambulances in these steep and wet conditions, so it was a nightmare of pain and suffering to extract the dead and wounded. Many Egyptians of the CTC froze to death in their summer-issue desert clothing as the British staff had failed to provide them with suitable clothing for the cold conditions. When their camels or horses couldn't move and blocked the dangerously slippery and narrow tracks, the soldiers had no other option than to push them over the cliff. The animals landed on the rocks below, spilling their vital burdens. And all the while, they were machine-gunned, deafened and blown apart by exploding artillery shells and surrounding rifle fire

Figure 22: Ancient Roman amphitheatre in Amman where modern classic and rock concerts are now conducted.

coming from hidden positions. The confusion was such that soldiers got separated from their units and their horses, some ended up with other units, some were captured.

Trooper Donald Black tells of one excursion where he and his mate Smithy had to recover horses lost in the battle:

> With some 18 horses gathered, we look for somewhere to camp the night. Breaking open a bale of tibbin and add some barley, we feed the horses, the first decent feed they have had for days. As for us, we have nothing … for seven days we have been constantly wet … shivering with cold and miserably hungry, we commence digging a circular trench with which to erect a shelter of sorts, using bags of horse feed … over the top and draping down the gap in the sides we place a blanket. Into this crude structure we crawl and lying on the damp earth try to sleep … our house is too short to permit stretching full length … some time in the morning our shelter came toppling in on us … I remained there with my head pushed into the mud until Smithy could extricate me … one of the horses had broken loose and his nibbling at the bags had caused them to fall … struggling out we sat huddled with our backs against the bags and our behinds in the mud, cursing the elements, the war and everything that irritated us … as day appeared and in mounting, Smithy pulled his horse down on himself … the horses had so felt the night chill they were too numb to support us and we walked them for about half an hour to restore their circulation.[197]

On a brighter occasion, Black found himself escorting five wounded horse-mounted mates back to the clearing station. He handed over his charges and gathered the horses for the return ride, then:

> Passing a section of the light car patrol my nostrils were assailed by the tantalising odour of cooking food. It brings me to a halt. The little we have had for many days has been scrappy and insufficient … the cook asks if I am hungry … this is too great a suggestion to pass and

---

[197] Black, Tpr D., Red Dust, pp. 88-9.

assuring myself they have plenty I eat my fill without fear of robbing others. It is only bully beef fried with onions but it tastes better than that.[198]

At the cook's suggestion, that it was too dangerous to return to his unit in the dark, Black stayed the night in comfort and enjoyed a hearty breakfast before his return with the horses.

Overall, the result of this assault was bad. But instead of calling it a failed attack the British called it a 'raid', as if it meant little. Unsuccessful British officers were making a mastery of self-deceit while refusing accountability. Smarter observers knew the reality and where fault lay. Reputations were tarnished.

To compound the disaster, the civil population of Es Salt, which had supported the EEF troops, now fled for their lives to avoid Turkish retribution but some didn't make it and were slaughtered. To help people escape,

> deeply moved, the Light Horsemen and Cameleers lifted exhausted women onto their saddles and rode through sleet and snow with children sleeping in their arms.[199]

We often see images of old and modern soldiers who, far from being merciless, are overcome with grief for the civil population, especially the women and children, the innocents in the mayhem.

In this raid, there were 1,350 casualties of whom half were Anzacs, resulting from the advance notice given to the Turks, the foul weather and inadequate air and artillery support. While over 1,000 Turkish and German prisoners were captured, the EEF troops were greatly pissed off with the conduct of this raid. To add to the cameleers' horror as they took prisoners,

> a number of Turkish and German prisoners had yelled for quarter; several of the [enemy] officers treacherously grabbed rifles they had thrown down and fired on the backs of the men who had spared their lives. In this way Captain Newson and two of the three officers were

---

[198] Black, p. 90.

[199] Hill, p. 144.

killed ... shot by an Austrian or German officer who had a moment before put his hands up in surrender.[200]

After twelve days of this hell, on 2 April a withdrawal was ordered; the troops withdrew to the west of the Jordan. Johnny Turk, on the other hand, enjoyed quite a morale boost after a string of earlier defeats.

In war, sometimes good luck is with you and sometimes it isn't. And poor decisions by commanders can mean win or lose, live or die. Allenby would make a second attempt.

As a precursor to Allenby's second raid, so-called again because it too was a bloody disaster, the Germans had launched a major offensive in France that had sent the British Army hurtling into retreat, with 38,000 casualties or prisoners in the first day. This coincided with the first raid on 21 March. Allenby was told to create an 'active defence' and send his prized infantry and artillery back to France. Over the next three months Allenby sent 60,000 experienced troops back to France. Luckily, or unluckily for Allenby, they were replaced by more untrained Indian infantry and cavalry over the next six months. Thus:

> Allenby, no longer able to attack in strength, was faced by the imminent task of reorganising his force and training masses of his Indian troops whose skills were often basic or even non-existent and whose inexperienced British officers were, for the most part, unable to speak their soldiers' language.[201]

Not only had the German offensive smashed the British in France, it unwittingly succeeded in major disruptions to the force in Palestine that now had to wait until autumn before being hurled against the Turks and Germans. Naturally, another reorganisation was needed. Chauvel's DMC lost the British Yeomanry, which was now a competent force, and it was replaced by untrained Indian cavalry.

---

[200] Langley and Langley, p. 119.

[201] Hill, p. 144

The Anzacs were about to be hit with two unsuspected and calamitous events. The first was the second raid across the Jordan from which they just survived, through good luck more than good planning; second, a hellish existence in a Jordan Valley summer that a non-indigenous man could not have imagined.

For this next attempt, Allenby's orders were for the infantry of General Chetwode to cross the Jordan River towards the Hejaz Railway. This they did on 18 April, with great show and bombardment of Turkish lines. This certainly got the attention of General von Sanders, who hastily reinforced his Turkish and German troops. What happened next is inexplicable.

On 20 April, Allenby withdrew Chetwode's infantry, which had made a spectacular show and had given notice to the Turks of a major action. He ordered Chauvel and the DMC to prepare to advance on Es Salt and Amman, then be prepared to move rapidly north towards Deraa in preparation for a final move towards Damascus.

He also ordered Feisal's NAA to attack Ma'an railway station in the south in coordination with Chauvel's attack. The main aim was to cut the railway and prevent Turkish reinforcements moving from the garrison there to help their mates in Amman. A secondary aim was to prevent troops from the bigger Medina garrison coming north in support. It would also prevent the troops in Amman from descending upon the Arab Army.

But impetuous Arabs, emboldened by earlier minor skirmishes, knew better. A handful of second-level commanders howled at Feisal and his army commander Ja'afr al-Askari, demanding to attack immediately. Until then, Arabs had shown a dislike for full-on frontal attacks, preferring guerrilla-style hit-and-run with a lower likelihood of death or injury (a good strategy). This time, however, this vastly over confident group caught Feisal and Askari in moments of weakness and strategic folly and they attacked the week before Chauvel. Turkish troops in Amman could not be prevented from assisting their troops

'Raids' Across the Jordan

in Ma'an. No surprise: the Arabs suffered heavy casualties and were driven off after a couple of days of military madness. The final result was that the railway was not effectively cut so support and supplies still got through to the Ma'an garrison.

Next, the Germans and Turks were prepared and ready for the assault on Es Salt and Amman that they had been forewarned was coming. However, von Sanders states, 'The preparations for the attack were made secretly and so skillfully, that their largest and last part was not discovered by our fliers or by observation from the ground'.[202] Nevertheless, von Sanders moved in large forces from west of the Jordan River across slightly submerged and hidden bridges, totally unsuspected by Allenby's intelligence staff. More came from Deraa in the north by rail.

The steep mountains, rocky gullies and narrow tracks once more prevented British wheeled artillery, supply wagons and ambulances being moved. The mounted men had to dismount and lead their horses and camels without infantry support. Chauvel was incredulous, dumbfounded perhaps. Even more so when Allenby, having parlayed with a small group of tribal envoys, directed that the local Bedouin tribe of the Beni Sakr would assist the DMC, believing they could link up with the NAA as they moved north. The unreliable Bedouin indicated they could help but no later than 4 May as, after that, they had to move camp to find seasonal harvests. Allenby had originally planned a mid-May assault. Had Allenby waited until then he could have used General Barrow's more reliable Indian Cavalry Division.

On top of this, Chauvel knew his logistics of supply and transport in this terrain were uncertain and any operation would be vulnerable to failure given the well-established position of the Turks and their ability to resupply and reinforce via the Hejaz Railway. He knew he could get to the top of the escarpment and Es Salt but getting to Amman, another fifteen miles of steep, scanty tracks, was very risky. In fact, they never

---

[202] Von Sanders, p. 221.

did get there. Allenby accepted this so gave no fixed time for the advance beyond Es Salt; which provided Chauvel with some relief but his trepidation remained.

With one eye on the Beni Sakr's promises and timeframe, Chauvel began his attack on 30 April. Anzacs and Indians performed as planned; but no Beni Sakr. Gullett notes: 'the Arabs may be dismissed at once. With their customary caution and fear of the Turks ... they withheld their support'.

Colonel Keogh reports:

> It came out later that the envoys represented only a small sub-tribe, and when they got back from the conference [with Allenby] they found that all their fighting men had gone south to join the Hedjazis [NAA].[203]

The British official historian, Cyril Falls noted

> that the undue reliance placed on the Beni Sakr came about because Lawrence was away on a mission and so GHQ 'made a mistake' in relying on the Beni Sakr.

He goes on:

> Jeremy Wilson, Lawrence's official biographer pointed out 'When Lawrence reached GHQ on 2nd May he learned to his astonishment that General Bols, Allenby's Chief of Staff, had just launched a further attack on Es Salt ... prompted by the Beni Sakr envoys ... this was a serious error on his part.[204]

Lawrence knew the Arab psyche, an understanding he could use to influence their thoughts and actions. He could also advise non-Arab tacticians about what they could and could not rely on. His presence in this instance was sadly missed.

The fighting raged for days. The airmen didn't notice the movement of Turkish and German forces from the west as they moved at night. Resupply was so difficult in those mountains and gullies that the innovative Colonel Downes, the Director of Medical Services for the DMC (and whom Chauvel had fought to have so long and hard with Allenby) arranged for

---

[203] Keogh, p. 221.

[204] Hughes, p. 85.

aircraft to drop medical supplies, the first time this was done in warfare. Unfortunately, the evacuation of dead and wounded was dreadfully difficult by horse or camel, whether walking or carried. A most difficult decision for soldiers is to leave a wounded mate or stay with him and risk one's own death or capture; the choice between sentiment and sanity. In those devilish conditions with the enemy all around and confusion running high,

> it was here that I heard, for the only time ever, of the pact between the troopers in the retreat, namely, if one was too badly wounded to ride, with the Turks close behind, he was to be left for capture. A tough bitter decision for light horsemen to make ... to leave their mates ... I do not know if this ever happened but saw a couple of badly wounded men brought in on the one horse by their mates.[205]

What a stinking choice for four good mates in a light horse section, what to do if one or two couldn't fend for themselves? Maybe worse was one experience of Trooper Donald Black on finding a broken body:

> The ground is saturated with blood from a gaping hole in his side. The leg too is shattered at the hip and the arm on the same side broken ... life for him will be an agony until he dies, as die he must ... I am undecided what to do ... with no hope of eventual recovery ... the machine-gun finds us and fires a burst like angry hornets ... one thought occupies me, a humane way out, but I am loath to adopt it ... meeting two stretcher bearers I ask them to come back with me ... they examine him, but I can see it in their faces if I didn't know it already ... they tell me it would take days of awful agony for him to die ... it was as if they read my thoughts but I cannot do it, the thought of shooting a man like this is abhorrent ... I look into their faces and fancy I see the same reluctance, no one speaks or moves ... one of them takes a hypodermic needle and fills it from a tube ... it seems a lot ... he replies that it is.[206]

---

[205] Hamilton, p. 115.

[206] Black, pp. 81-2.

At one stage Brigadier Grant's 4th ALH Brigade was surrounded, attacked and looking done for:

> Our sector is quiet tonight ... just as well as we are too few to hold it ... we are too sparsely scattered to hold it should an attack come ... the horses have been tied head to head to release some of the horse holders, for casualties have greatly thinned our ranks ... we keep firing our rifles, but it is only make believe, to create a suggestion for the Turks benefit, that we are more numerous than we actually are.[207]

By some mistake, Turkish HQ withdrew their 2nd Regiment and moved them north, relieving pressure on the 4th Brigade, the heroes of Beersheba. Just the same, nine of the twelve artillery guns, gun ammunition and an armoured car were captured. There is no greater dishonour to artillerymen than to lose their guns, and that became the subject of an enquiry later. Even the 4th's Field Ambulance copped it and, as acknowledged in the *Australian Medical Service Official History*, 'these were the most difficult and dangerous marches undertaken by medical units in the whole campaign'. Warrant Officer Hamilton of 4th Light Horse Regt Field Ambulance recorded:

> Wednesday 1st May 1918 was one of the most memorable days of my life ... our ambulance had gone, captured ... we lost four of our own ambulances, all our hospital equipment that included our precious medical and surgical panniers, with all surgical instruments. The roll call on 2nd May showed 12 of our men missing believed captured.[208]

Had the 2nd Turkish Regiment stayed there, it seems of little doubt that the 4th ALH Brigade and its field ambulance would have been annihilated: luck stuck with the Anzacs that day, albeit with losses and casualties.

When the Turks took those twelve medical prisoners,

---

[207] Black, pp. 82-3.
[208] Hamilton, pp. 113-4.

the first thing [they] did was to make them take off their boots. This served a double purpose. In the first place 12 lucky Turks got a good pair of boots ... Turkish footwear was always shocking and often all they had were rags wrapped around their feet. Second, it prevented them running away ... on the hard rocky ground their feet were cut to pieces.[209]

Even non-combatants suffered.

During the five days of this raid, the aircraft of the AFC and RFC were active. They bombed, strafed, fought off the German planes that approached and took reconnaissance photos. The Brisfits continued to show their dominance over German aircraft and pilots but the German Taubes created some havoc, even bombing the field ambulances and CCSs.

But not everyone found it tough going. Lieutenant Clive Conrick, an observer in No. 1 Squadron, diarised on Wednesday 1 May that the morning patrol had gone out and was an hour overdue returning and that his bombing raid had been delayed by bad weather. In this case:

> At 1030 Beaton and I went by car to Jaffa. We had dinner at the Jerusalem Hotel. The meal was very enjoyable. We both went around the markets and canteens where we did some shopping. In the afternoon we bought some wine for the mess, which cost us £3.0.0. Later in the day we were invited to the home of some Russian friends. Most of the lady folk present were amused at us because we were all wearing shorts.[210]

An airman's life may sound more leisurely than that of the ground warrior but a crash, forced landing and being captured by Turks or Arabs quickly took away that gloss:

> Lieutenant J.K. Curwen-Walker and Corporal N.P.B. Jensen took off at dawn this morning. The machine stalled at 500 feet, got into a spin and crashed in front of us. The Bristol Fighter burst into flames on impact and both the crew were killed. A sad beginning to the

---

[209] Hamilton, p. 117.
[210] Conrick, Lt C., The Flying Carpet Men, personal diary, entry 1 May 1918.

Desert Anzacs

day. Captain Alan Brown and Finlay with Oxenham and Leith escorting, took off immediately afterwards on the dawn patrol.[211]

Conrick's entry the next day laments:

> Two Huns flew over this morning. They dropped letters from our missing airmen ... It was a sad story that they told for Captain Rutherford had both his petrol tanks punctured by machine gun fire and was forced to land. While Rutherford and McElligott were burning their aircraft, Fred Haig landed beside them ... when he tried to take off again with his two extra passengers, one wheel collapsed before they were airborne and his Bristol Fighter stood on its nose. There was nothing the four men could do but burn their machine and surrender to Circassian Cavalry who galloped up and surrounded them ... handed over to the German Flying Corps and sent to a prisoner of war camp. This makes a total of twelve of our Squadron officers in enemy hands.

As war is, good follows bad follows good. Conrick also recorded on 2 May:

> McGinnes [one of the founders of Qantas] and Hawley shot down a Hun 2-seater this morning ... McGinnes narrowly avoiding a collision with the falling aircraft ... Lieutenants H.S.R. Maughan and Hudson Fysh [another Qantas founder] reported after their noon reconnaissance that Turks were rushing reinforcements ... and ... at Amman there were at least 500 infantry and 100 rolling stock at the Station, with another train just pulling in ... nearby a new hospital had been erected and nearly 200 horse-wagons as well as a large number of cavalry and infantry were dispersed.[212]

This invaluable intelligence helps commanders make valid decisions on how to conduct their battle plans: to attack or withdraw, to fight now or flee to fight another day. Allenby at last sent in General Shea's infantry to assist on 3rd May: too late. On 4 May, Dickie William's squadron loaded up with bombs,

[211] Conrick, 1 May.
[212] Conrick, 2 May.

although their role was photo reconnaissance, and gave what they could to the Turks.

Also, on 4 May, Allenby went forwards to see for himself. He met Chauvel who simply said they had no chance in those conditions against such a well-entrenched and prepared enemy that had been reinforced and resupplied. Allenby concurred and Chauvel ordered a withdrawal on 5 May. In the withdrawal, just as was the case in the first raid, civilians were in flight. Once more, the horsemen gave their saddles to weary women and children.

Just like the EEF troops, so too were the Turks and Germans exhausted. Von Sanders reports, 'Our troops on the fourth day of heavy fighting were approaching the limit of their endurance'. As the British finally withdrew, 'Unfortunately the troops no longer had enough strength to complete the success.' He continues, 'At any rate, we had succeeded once more in preventing the enemy from gaining a foothold in the East Jordan section'.[213] Von Sanders had fought a good fight and Allenby had made critical errors. The last of the 4th ALH Brigade crossed the Jordan at midnight on 5 May. The second battle, far more fierce than the first and unworthy of the title 'raid,' was over.

In the aftermath, Allenby criticised Grant of the 4th and the gunners who lost their guns as well as Chauvel for not reinforcing Grant's flank. An enquiry showed that had Allenby waited for Barrow's Indian Division instead of relying on the ever-wavering tribesmen, Chauvel would have had sufficient troops to provide adequate protection for Grant and the guns, and the whole operation may have had a different result.

Chauvel expressed regret for parts of the operation and that it had failed. 'Failure be damned! It has been a great success!' was Allenby's mysterious retort that he offered to Chauvel some time later.

Perhaps the success that Allenby referred to, was that these two disastrous raids provided evidence to von Sanders that the

---

[213] Von Sanders, p. 230.

future major assault would come from east of the Jordan, along the Hejaz Railway. Or perhaps this was just another case of 'oh, how men deceive themselves'. Or, as Colonel Keogh suggested, 'perhaps the basic reason [for the British defeat] is that they all took 'Old Jacko' a bit too cheaply.[214] In any case, many good lives were lost.

It seems Allenby was somewhat surprised by this outcome. He failed to see its risky nature even after the event. He, was, however, unusually quiet in conference with generals Shea, Chauvel and his staff in the wash-up. Yet, the survivors were given a lame story that they had succeeded, but they weren't deceived:

> The High Command's explanation is that we only came up here to gain information, which having been successfully obtained, there is no longer need for our presence. If that be true, it has been exceedingly costly. We feel very sad as we ride along, everybody is leading spare horses, some two and three. Each empty saddle means a comrade passed on, but we hope some day to avenge all this. Should we come again we will repay'.[215]

Considering these events twenty years later, General Chetwode wrote to then General Wavell, 'These two expeditions of Allenby's across the Jordan were the stupidest things he ever did, I always thought, and very risky'.[216] On the other hand, Hill suggests they weren't necessarily stupid but were extremely risky. In most other respects, Allenby was a commander of intense skill and ability – which makes these raids all the more unfathomable and totally out of character. And, as he left no memoirs or diary, nor seems to have confided in his subordinate commanders, who themselves left no written explanation, his thinking is harder to explain than the Bermuda Triangle.

There was a further consequence. The Arab tribes of the region that were not fans of Feisal or the Hejaz Arabs declined

---

[214] Keogh, p. 228.
[215] Black, pp. 90-1.
[216] Hill, p. 152.

'Raids' Across the Jordan

to join the Arab Revolt. They saw that the Turks were still in Medina and Ma'an and had defeated the British twice in raids across the Jordan. Some Arabs saw no tangible signs of victory by Feisal or the other Arab armies. They once more sat on the sideline. At least they didn't fight against the British.

It is likely that Allenby had wanted to create the notion that his major actions in the future would be in the east of the Jordan. The first raid partly showed this. Also at this time, as Colonel Keogh reported, 'an immediate seizure of the plateau would deprive the Turks of great crops of ripening wheat and barley, therefore, their survival.[217] Allenby's actions had some effect, as General von Sanders acknowledged after the second failed raid:

> It was not impossible that a third enemy attempt to break through our lines and into the East Jordan section was contemplated in order to make contact with the Arabs fighting there.[218]

These raids across the Jordan were a setback. But Allenby had succeeded at Beersheba, Gaza, Jerusalem, Bethlehem, Nazareth, Jericho. He'd harmonised the mixed nationalities of the soldiery with the Egyptians of the Labour Corps and CTC and the engineers driving the railway and roadways.

As summer came, more Americans came into France relieving the need for more British manpower there. This allowed Lloyd George and his easterners to get serious about driving the Turks out of Palestine and delivering Syria to them. 'Airplanes, anti-aircraft guns, food and uniforms all arrived in abundance, making the lives and combat of the individual soldiers somewhat easier'.[219] Confidence grew among the EEF. Training had a new purpose. Reports from new prisoners and spies showed that the Turks were a depleted and sorry lot, adding more confidence to the EEF. Air domination was now almost complete and only the formidable Jordan Valley seemed to be the unwanted boil on a horseman's bum.

---

[217] Keogh, p. 219.

[218] Von Sanders, p. 233.

[219] Woodfin, p. 123.

Part Four

Armaggedon

17: Who Wants Palestine?

There are three sides to every story – yours, mine and the truth.
– adapted from various sources

Now politics intervened in the eastern war. Save the world from evil, or save the world for the empire?

> Britain fought the Palestine campaign in order to maintain the empire. It was necessary to protect the route to India via the Suez Canal by creating a chain of contiguous territory under British influence, and this required some form of control over the Middle East. Better if it could be achieved through indirect and peaceful means, rather than military conflict. The Arabs of the Hejaz represented the indirect means of extending British rule. Support for the Hashemites would be a less expensive way of looking after British interests and in a world altered by Wilsonian[220] ideas of self-determination, local allies were a less obtrusive means of control. Zionism would fulfill the same purpose in Palestine. To make the scheme work, the EEF had to conquer the Levant and instill Zionism in Palestine and the Hashemites in Syria. By doing this Britain could 'unmake' the Sykes-Picot Agreement of 1916 that would internationalise Palestine and allow French control of the area from Beirut to Mosul.[221]

---

[220] US President Woodrow Wilson forced his view on nations that had fought throughout the war that peoples of the former Ottoman Empire should be allowed to determine their own future rather than have the League of Nations determine it. Britain and France preferred to be in control of the Middle East oil and trade routes. They also suggested that the 'natives' of those areas were not ready for self-determination or self-rule and should be directed to accept the tutelage of more developed nations; themselves.

[221] Hughes, p. 89. The British Government led by Lloyd George wanted to exclude France from the Middle East if they could, so politics became devious. France was equally demanding.

Desert Anzacs

Leopold Avery[222] wrote that 'Britain's dilemma was twofold: firstly, how to win the war; secondly, how to ensure post-war national and imperial security'[223] – read wealth and control.

The French Government wanted Syria and Lebanon, where for decades it had developed schools, universities and religious centres and liberated vast archaeological treasures. It wanted a Mediterranean seaboard for protection against Russian naval growth and a gateway for the removal of more spoils from its regional archaeology projects. It clearly supported the view that self-determination was absurd and only French tutelage could safely and effectively guide natives.

The Arabs, underdeveloped but ambitious for self-determination, wanted everywhere: from the bottom of the Arabian Peninsula to the bottom of Turkey; from Lebanon's coast to Iraq. Some wanted an independent Arab nation and Caliphate with Sharif Hussein as King of the Arabs and Caliph. But not all Arabs wanted Hussein as their king. In fact, few did; tribal cohesion being non-existent. Tribal Bedouins wanted gold, weapons, food and their nomadic culture, and cared nothing for an Arab nation.

Allenby wanted to eradicate the Ottoman and German forces. Chauvel's Anzacs wanted to win the war, go home and likely didn't care about imperial anything.

By late May 1918, the raids across the Jordan were military failures but served Allenby's strategic ploy of influencing von Sanders to deploy one-third of his force in the east. Allenby knew the wet season ran from November to May when the rains would make movement difficult. He chose mid-September to take control of the Levant.

He had much to do: replace his experienced soldiers with <u>untrained ones</u>; deploy his force to pretend they weren't where

---

[222] Avery was a Conservative MP in the Lloyd George government who helped draft the *Balfour Declaration* and later assisted the formation of the Jewish Legion for the British Army in Palestine. He dismissed President Wilson's self-determination concept and the vote of the League of Nations as absurd, on the basis 'not all nations are equal' (having varied levels of economic and social development).

[223] Hughes, p. 89.

he really wanted them; reorganise his troops; resupply with new equipment; build up vast supplies of stores; gain control of the air; prepare his medical plan; feed everyone now and in the final attack; and reinforce and resupply during an attack that would be fast. Over 300 miles of rail, roads, water pipelines and airfields had to be prepared.

Commanders excel when they understand logistics. Excellent commanders become victorious commanders when they get their logistic and battle plans right. Allenby knew this. Chauvel knew this, so he, too, was thinking ahead.

The Mediterranean port of Jaffa and the nearby city of Ludd (Lydia of the Bible) became enormous bases where stores were built up. Rail and road links were built and improved. Extensive water pipelines were improved to support troop build-ups. Weapons, armoured cars, artillery guns, ammunition, food, horse equipment, aircraft and spare parts, medical and veterinary supplies; all were distributed and stockpiled. Soldiers beamed with confidence that 'something big' was going to happen. Maybe an end to this damn war.

The War Office asked Allenby what he was going to do. His reply: to control the area from Tiberias on the Sea of Galilee to Haifa on the Mediterranean coast and to the Jordan Valley and Jordan River in the east. This would give him two paths for a final invasion and the destruction of the Ottomans. He planned his main attack to be up the coast while the Jordan front would be his feint.

Allenby and Chauvel knew that possession of the Jordan Valley was essential to the success of any future offensive. If Allenby abandoned the valley his only avenue for advance would be the Plain of Sharon along the Mediterranean coast. With the Turks already concentrated with two armies against him in the west, the prospect of his breaking through would be reduced. He therefore decided that the valley must be occupied and the river controlled from the north of the Dead Sea. Chauvel was to do everything possible to persuade von Sanders that his horsemen were the main thrust.

Allenby gave Chauvel the choice between actually remaining in the valley and its coming horrific summer, or withdrawing to the more habitable heights behind Jericho with the understanding that when the moment came, he must win back the valley then cross the river again – an option easier up front but mighty hard at the end.

Locals warned that during the summer, the Jordan Valley was uninhabitable and even the nomad Arabs fled to the hills and every resident of Jericho, except the very poor, evacuated the village as soon as the winter season was over. For white soldiers it was unthinkable. Chauvel, thinking the unthinkable of remaining in the valley, wrote:

> There were three reasons why it should be held. The first, because the road from the Turkish railway at Amman, crossing the Jordan at Ghoraniye, was always a serious menace to our right flank; the second, because it would be necessary to retake it before the advance in the spring and it was considered that it would be less costly in lives to hold it; and the third, because it was desired to hoodwink the enemy by the display of a large force and constant activity on that flank.[224]

He prayed his troops could survive the summer to fight in the autumn. In addition:

> In the jumble of hills overlooking the Jericho plain, which was the only alternative, there was neither space nor water for a large body of cavalry. It was, therefore, decided to hold the Jordan Valley and do what we could to combat disease. Though our losses from malaria were considerable, the heat intense, and the dust worse than our troops had hitherto experienced, the ultimate results more than justified this decision. Captured papers revealed that von Sanders and the Turkish High Command assumed that, in whatever part of our line the Anzacs were in evidence, it was from there we would be expected to strike.[225]

---

[224] Gullett, p. 639.

[225] Gullett, p. 641.

Who Wants Palestine?

Bait taken once more.

Preparations were made for the occupation of the river flanks. Brigadier Wilson's 3rd ALH Brigade moved onto both sides of the Jordan and began to work hard on its defences of trenches and wire entanglements. Brigadier Cox's 1st ALH Brigade was deployed to the west of the river. Chaytor's force spread south and east along the Jordan to the tip of the Dead Sea while other Anzac and Indian brigades spread throughout the valley. They all had major trench and earthworks to dig in the wilds of the valley. Such earthwork was unknown to horsemen. They weren't supposed to be in one place long enough.

Talking about occupying a place is one thing but living in it is something else. This Jordan Valley was about to throw three months of purgatory at the Anzacs that would darken their memories for years ahead.

> Everything about this valley suggests that Nature was in humorous vein when she conceived it. Conceived in humour perhaps, and finished with spite. For more than a suggestion of spite has been added to it. It would seem what God commenced, the Devil completed.[226]

So the DMC moved into the Jordan Valley. And what a perfect time to reorganise once more: new Indians, new equipment, redundant camels, new ground and nothing new about the delay.

Yes, it was the end of the road for the camels. The Es Salt/Amman raids marked the end of the fighting career of the ICC. This force, under the quiet, capable leadership of Brigadier Smith VC at Romani, Magdhaba and Rafa, in two Gaza engagements and at Jerusalem and Jericho had brilliantly justified its retention. On the firmer ground of Palestine, the horses and the wheeled and motorised wagons outpaced the camels. Now, in the hills, mountains, valleys and ravines, the camels' splayfooted pads were slow and unstable. The men would be of greater value mounted on horses. So decided Chauvel and Allenby.

---

[226] Black, p. 97.

It seems Anzacs will find merriment in most things, always being ones to look for the bright side of life:

> In June 1918, a mock funeral was held for the disbanded Camel Corps. The 'corpse' was a camel saddle, covered with a flag and carried on a stretcher to outside the 'chapel' (my tent) where it was lowered into its grave. A body (of Cameleers) made of comrades that came from outback Queensland, paddocks of New Zealand, stately homes of England, all classes, all types – welded together by an equally conglomerate mass – that incredible animal, the camel.[227]

A eulogy was given, words said and a guard of honour fired a volley. Soldiers who had experienced the death of mates and near-death of themselves and developed a bond with their dromedary friends were moved to emotion; yet, they understood the future and let go the past, some with a tear, all with gratitude.

The Australians of the camel brigade were put on horses, given swords and formed the 14th and 15th ALH Regiments. How hard could it be to move from a camel's hump to a horse's rump? Not only moving from a camel to a horse but those new to horses also had to familiarise themselves with new weapons; out were their bayonets and in were swords of cavalry. The temporary training officer was Brigadier C.L. Gregory, a British cavalryman from the Indian Army. Being British, Gregory was initially doubted. He quickly demonstrated skill and understanding and 'was an expert trainer who became popular and respected. He brought the 14th and 15th regiments quickly up to scratch in ten weeks and was cheered on his departure'.[228] One of the finer British officers.

To the 14th and 15th was added a French colonial regiment of Spahis and Chasseurs d'Afrique. The three regiments became the 5th ALH Brigade under the command of Brigadier George Macarthur-Onslow, promoted from command of the 7th ALH

---

[227] Langley and Langley, pp.145-6.
[228] Hill, p. 154.

Regiment. Onslow was a descendant of John Macarthur, famous as the founder of the merino sheep industry in Australia. He went to the new brigade with a brilliant record as a regimental leader:

> Although no deep student of tactics, he had, as an Australian countryman, a very shrewd sense of ground and was by instinct a dashing leader of horse. Fiery in temper, but gallant and generous in bearing, no light horse leader rode harder or straighter in action than George Onslow. His men of the 7th Regiment had trusted him implicitly and followed him blindly, and as a vanguard in serious operations they had no peers in Palestine.[229]

He became one more brilliant Anzac leader.

The New Zealanders of the Camel Corps became the machine gun squadron for 5th ALH Brigade. The Australians who had made up the camel field ambulance were mounted on horses and transferred to Macarthur-Onslow. The British battalion of the camel corps retained their camels and were sent to operate independently in the southern desert and support the Arab armies against the Hejaz Railway. And a bold request by Lawrence to Allenby saw 2,000 camels given to the Arabs as another reward for assisting the British cause.

Most of the now experienced and acclimatised British yeomanry of the DMC were lost to France. In their place arrived Indian lancers that had been wasted by doing not much in France. More were newly arrived from India with no battle experience. All turned out to be good horsemen, coincidentally most astride the well proven Australian walers:

> Several times their advance patrols had galloped down bodies of Turks and their terrifying use of the lance in those small engagements had a highly useful effect on Turkish nerves and morale.[230]

---

[229] Gullett, p. 650.

[230] Langley and Langley, p. 149.

They formed the 4th Cavalry Division under Major General Barrow, a capable and proven British officer. Lots of training was needed to go with their keenness in action that saw them dive into their preparations like hounds into a foxhole in an English meadow.

In addition, the 5th Cavalry Division was formed from two yeomanry leftover brigades and another Indian brigade, under Major General H.J. Macandrew, another British officer of the Indian Army.

Chauvel now had four mounted divisions with 30,000 horsemen from Britain, Australia, New Zealand, India and France and artillery from Britain and India, a truly imperial gathering under a colonial commander.

It took until August to settle this mix of nations, which left precious little time for the training they needed before the big offensive. They alone would have been enough to terrify the remnants of a worn-out Turkish Army. But there were more.

A couple of hundred thousand infantrymen would harrow the Turks. Once more Allenby lost his best troops to France and saw them replaced by Indians with mixed experience but ferocious keenness. Keen, yes, but with training desperately needed. Deployed in the west along the coast and in the centre on the plains with generals Chetwode and Bulfin and the remaining British infantry, they got their training.

Some say 'practice makes perfect'. But that's wrong. Only perfect practice makes perfect. When lives are the players and victory the trophy you have to get it more right than the enemy. The simple rule for winning in sport is to have more points on the board than the opposition at full time (except golf). So it is in war. Training, perfect practice, rehearsals, discipline, the right equipment, the right team support, exemplary logistics, accurate knowledge of the opposition, high morale and good healthcare, as well as the complete mastery of individual and collective tasks were essential. And a brilliant plan.

In doing this,

Who Wants Palestine?

the training of the Indians was pushed on at high pressure, and the two new Australian Light Horse regiments formed out of the Camel Brigade were given their horses and schooled afresh. Many of the men had originally been in the light horse, but perhaps half of them had been drawn from the infantry and of these many had never been accustomed to a horse.[231]

While all this was going on, Allenby didn't indulge overconfidence. He feared the Turks could receive reinforcements and hit him east of the Jordan. And as late as June the War Office, despite the arrival of American troops, was still taking his experienced infantry, artillery and cavalry to bolster their paranoia about losing France while sending him untrained Indians. His soldiers, however, were receiving new aeroplanes, guns, armoured cars, machine guns and ammunition. Forward airfields were built. The railway followed. Roads and vehicle convoys built up stores. The force had mixed battle experience and ability. But most of all, the men were motivated. Allenby still wouldn't be ready to advance before September. And a nasty fright would arrive in July.

Being the kind of guy he was, Allenby never faltered. He planned with total confidence. He instilled confidence of an overwhelming victory into his generals, who instilled it down the line to division and brigade commanders, who kicked it down to every soldier, whose confidence developed without bounds. The anticipation that something big was on the cards filled hearts with hopes of victory, like the first dinner with a new flame.

Even the Arabs got enthused. Of all people, the Beni Sakr began hitting the Turks around Amman. Tribes in the south attacked outposts and stopped rail movements until nothing was moving north or south along the Hejaz Railway. The Turks in Medina and Ma'an were now isolated and unsupported and couldn't come out to help.

---

[231] Gullett, p. 657.

War from the air had taken a new twist. Nothing moves a soldier more than the unexpected bombing or strafing from an aeroplane. The Germans had the advantage in the early years. From late 1917 the RFC started to receive their superior Brisfits. From early 1918, No. 1 Squadron received theirs. These planes turned the tables on the Germans for the rest of the war. The Brisfit was not only more powerful than the old BEs and RE8s, it was an aggressive weapon as well, with good performance and endurance:

> Ross Smith tested this endurance when on 30 January he flew one to 19,000 feet to test a new type of camera. Unfortunately, the fuel pump froze on them, which caused them to come down again.[232]

Captain Ross Smith DFC, MC was Commander 'C' Flight 1 Squadron AFC. Initially enlisted as a trooper in the 3rd ALH Regiment in Adelaide in August 1914, he served at Gallipoli, where he was promoted to regimental sergeant major, deserving of this meteoric rise in rank. He continued in the ALH and was commissioned as an officer, serving in a Machine Gun Squadron at the Battle of Romani in August 1916 where he first met and impressed General Chauvel. In July 1917, he responded to the call for volunteers to the AFC, where he became an observer in December before gaining his pilot wings around April 1918. Recognised as

> one of the great leaders and larrikins of 1 Squadron he was the highest scoring ace on the Palestine Front; one of only two pilots in No. 1 Squadron (along with A.W. Ellis) to destroy an enemy aeroplane in an aircraft other than a Bristol Fighter. By war's end he had been awarded the Distinguished Flying Cross (three times) and the Military Cross (twice).[233]

Generally, he was acknowledged to be the premier pilot in the AFC and the RFC on the Palestine front. After the war

---

[232] www.southsearepublic.org/2004_2002/people/aces/smithross.html, p. 5.

[233] http://www.southsearepublic.org/2004-2002/people/aces/smithross.html, p. 2.

he and his brother Keith entered and won an air race from England to Australia, for which they received a prize of £10,000 and a knighthood each. There is a statue to this great aviator in Creswell Gardens in Adelaide and their plane is on display at Adelaide airport. He'll reappear with some heroic and mischievous deeds.

In the meantime, Allenby and Chauvel had to keep the men and the animals alive and well to run their campaign. The medics and vets came to the fore.

The aim of medics is simple: get as many blokes home as possible. Chauvel well knew the importance of hygiene and the health of his troops. With Colonel Downes, his chief medical officer,

> the medical service was developed to meet the requirements of extensive operations. By August, Casualty Clearing Stations [CCS] were distributed at Ludd, Jaffa and Jerusalem (including a hospital). Two advanced depots of medical stores were at Ludd and one at Jerusalem. On the lines of communication between Kantara [on the Suez Canal] and Ludd accommodation in stationary and general hospitals could provide for 10,000 British and 5,000 Indian sick and wounded.[234]

In addition, evacuation arrangements were made using motor ambulances and hospital trains. Welfare was now carefully planned. Welfare and morale were clearly more important now than to the Murray-led staff officers.

The medical service of the EEF had other considerations. General Bulfin's 20th Corps in the hills between the Jordan River and the coast had only the heat to contend and fewer nasty bugs, critters and their diseases. But for Chetwode's 21st Corps on the coastal plain and Chauvel's DMC, it was altogether different 'to meet the effects of prolonged residence in one of the most malarious districts in the world'.[235] In addition, the DMC had to survive in horrendous terrain. That required special services

---

[234] Downes, p. 696. There were more British and dominion soldiers than Indians.

[235] Gullett., p. 697.

Desert Anzacs

if the soldiers were not only to survive, but to maintain their stamina and fighting condition, as well as write home in happy fashion using their usual humour and candour.

It is reported that, hundreds of years earlier, one crusade failed due to the reduction of the troops by malaria and, that Napoleon also attributed the failure of his Syria campaign to malaria.[236] The greatest threat in these months of limited fighting in the Jordan Valley 'was the endemic prevalence of malaria' where 'the greatest problem faced by the medical services in the whole campaign was its prevention'.[237]

Prevention and treatment were instituted. Every man in the EEF was given personal responsibility for his own health measures. But some considered it effeminate to sleep in a mozzie net, wear gloves and veil, take pills and a non-effective repellant cream; others did whatever was needed to avoid being bitten. The medics and engineers attacked mosquito breeding grounds. Oil was sprayed over stagnant ponds to kill the mosquito larvae. Miles of channels were dug by the troops, in the sun and heat, adding to their discomfort, to widen the waterways and get the water flowing. Unfortunately, nobody told the mozzies on the Turkish side to stay over there and squadrons of the untreated beasts invaded the EEF side.

For health and tactical reasons it was decided 'to hold the valley lightly, and give each of the four divisions in turn only a short tour of duty – from four to six weeks'.[238] The released troops were sent to Jerusalem and Bethlehem Red Cross camps with tents, mattresses, games, books and comforts. Visits to the Empire Soldiers Club were greatly anticipated and enjoyed – happy days.

In the July to the mid-September period, only three to four per cent of the force was admitted to the Field Ambulance for disease. This number multiplied horrendously from mid-October into November once the offensive started and

---

[236] Gullett, p. 706.

[237] Gullett, p. 705.

[238] Gullett, p. 712.

prevention measures couldn't be maintained.

Then there was the heat and the dust. Night brought little relief from the heat and no relief from the dust. The men carried on their duties of patrolling, digging, wiring, caring for their horses and mosquito prevention work in temperatures above 40°C, sometimes as high as 50^0 C. At least they had more water for drinking and washing than they did in Sinai, so the heat was a discomfort rather than a health issue. But the dust was something else. 'On the stillest of days a single horseman would raise a dust cloud that completely obscured him and was slow in settling'.[239] Powdery like talcum, it covered a man's features, changing his appearance so mates couldn't recognise one another. A horseman without a horse could be mistaken for a fence post until he moved. It filled his gear so he couldn't keep rifle and kit clean, clothing rubbed grit against the body and, the hills towards Jerusalem were lost in a fog.

> Whirlwinds – huge, whirling columns moving slowly over the face of the valley. But it is the dust-cloud raised by the wind or transport that chokes and blinds the moving troops. Men in the rear of a mounted column will see nothing either side for hours, and scent nothing all day but the dry, strong smell of powdered earth that parches them.[240]

Mealtime was a mix of food and dust. The dust was dirty, unlike the cleaner sandy dust of Sinai that only blew with the khamsin for a few days. This dust was infinite. And the monotony of bully beef, vegetables and bread so affected the men that the depressed troops had short tempers and had to apologise to those they had yelled at when they recovered their equilibrium. This was one bad, bad place.

Another affliction the horsemen faced was boils. Since the time of Moses the irritation caused by dirt and the dust was a breeding ground for boils around the neck, armpits and buttocks:

---

[239] Gullett, p. 702.

[240] Dinning, H., p. 63

> Boils are very painful and in bad cases can put a man out of action. The best treatment is to lance them, which gives immediate relief. But they must be 'ripe', to let out the pus. Otherwise hot foments until they are ready for the scalpel. One of the worst places is on the buttocks.
>
> For a light horseman to ride is at times impossible.[241]

There was also the fairly regular incursion of snakes, scorpions, spiders and local dogs to add to grief and swear jars.

In many cases, after hospitalisation, convalescence in Egypt was necessary. Here, even the carers and staff were horrified at the changes in the men:

> Matron M.A. Early of the Aotea New Zealand Convalescent Home, who received her countrymen, said that 'the poor boys used to come to us looking utterly broken and old – tremulous and shaky. It was indeed hard to see the woeful change in our sturdy, healthy-looking men'.[242]

The Anzacs had it a lot better than Johnny Turk, with no medical care and morale at rock bottom. 'Dressed in rags, Turks who captured British soldiers stripped them of clothing and food'.[243] Indeed, they stripped them of everything they could for their own use, being so deprived by their own officials and officers.

The horses got sick, shot and blown up. They needed good food to maintain their stamina and they got exhausted, just like their riders. In May, after the raids across the Jordan, 9th MVS treated and returned 110 horses to their units and evacuated 101 that had to be replaced by the rough riders of the Remount Units.

Supplementing the local grain with fodder suitable for Australian walers whose stamina and endurance was way superior to local Arab ponies was a constant 'search and recover' activity for the vet staffs and supply officers of the

---

[241] Hamilton, p. 132.

[242] Woodfin, p. 130.

[243] Woodfin, p.130

DMC. Fortunately, the purer water of Palestine compared to that of Sinai meant less disease and debility for the horseman's best friend.

But the Jordan Valley was kind to no one:

> Horses unable to travel to Jerusalem had to be held under most adverse conditions, in a dust laden atmosphere, under a scorching sun, some hundreds of feet below sea level and in a place where every type of flying and crawling insect was well represented.[244]

The vets and the rough riders were a vital part of the team that kept the horsemen on horses, the DMC mobile, and Chauvel's force combative. This all meant Allenby had an army. It also meant the vets, in preparation for the coming offensive, had to bulk up their medical equipment, medications, bandages and transport and prepare their staff for the fighting that even they would be part of. Shot and shell near the front line was not something the typical vet experienced at home so mental preparation for this was needed.

---

[244] Anon, unit history, handwritten notes, 9th MVS attached to 4th ALH Bde, undated.

18: Crush the Enemy

If you ever dream of beating me, you'd better wake up and apologize.
– Muhammed Ali

Morale and welfare were never concerns for the Turkish Pashas. Men were often taken from family, given a rifle and some rags to wear, poorly fed, unpaid, and put into a no-win situation, creating desertions a plenty.

> The poor old Turkish soldier lacked almost everything, especially good food and boots. His animals were so under-nourished that they could not be relied upon to haul his guns and, while the soldiers could escape their misery by deserting, the draught animals could only die. This they did in the hundreds.[245]

For Allenby and Chauvel's soldiers,

> they were quick to recognise that they were opposed to an enemy whose spirit was low and whose interest in the struggle was dissipating. Nearly all the intelligence reports during the summer made cheery reading for the British. The quarrels among the enemy's High Command became more acute; the strain on their narrow single gauge railway was excessive; supplies were short and irregular.[246]

In reality, had British intelligence been that knowledgeable and commanders interpretative, perhaps the final attack could have been conducted a year earlier and, in doing so, ended the war earlier.

The Turkish soldiers had been told of outstanding German

---

[245] Hill, p. 163.
[246] Gullett, p. 655.

victories in France and seen the raids east of the Jordan go their way. But they knew they had been pushed back from the Suez to near the end of Palestine by crazed horsemen. And not all the Turks could be fed nonsense:

> The stolid Turkish soldier paid little heed to German reports. He was too well accustomed to seeing his own victories exaggerated and his failures explained away. His religious fanaticism had long since burned itself out. When his trenches were boldly raided he offered but feeble resistance and seemed pleased to be taken prisoner. More stimulating to the British was the steady and growing stream of deserters during the summer. These men arriving each night were in no mood to be secretive and while their desertion was evidence of the low spirit of the Turkish armies, they also brought much exact and valuable information.[247]

Others stayed to fight on. Poor ol' Johnny Turk had put up a good fight, but was seriously let down by his pashas and corrupt officers. The tenacious German military mission of von Sanders could not overcome the immaturity and reckless inexperience of the pashas. In odd incidents it seems some Turks took their combat tasks light-heartedly:

> During a battle in the summer of 1918, a large group of Arab civilians gathered on a distant ridge to watch the 'free show' where a group of New Zealanders were enduring an artillery barrage from Turkish guns. To everyone's astonishment, the Turkish and Austrian gunners shifted their fire off the New Zealand trenches and began shelling the civilian onlookers. The Kiwis watched 'with some satisfaction' as the panicked Arabs scattered.[248]

General von Sanders was beside himself, aghast that the Pashas then did a 'British War Office'. Experienced Turkish troops were withdrawn to the Caucasus and Mesopotamia, replaced with 'Turkish battalions with hardly any efficient

---
[247] Gullett, p. 656.
[248] Woodfin, p. 131.

officers, and devoid of training, so that the army commanders preferred to distribute them as replacements'.[249]

He described the state of the three Turkish armies in Palestine for which he was responsible as army commander:

> The number of Turkish deserters is higher today than that of the men under arms. The clothing of my army is so bad that many officers are wearing ragged uniforms and even battalion commanders have to wear sandals instead of boots. The troops today are undernourished, very poorly clothed and wretchedly shod. The traffic changes of the railroad with its very limited equipment would make it impossible to supply the remaining troops with war material and subsistence. It becomes my duty to state that the present state of the Turkish army does in no way permit it of far reaching operations.[250]

But that wasn't the worst for von Sanders.

Enver Pasha still pulled the strings of political power over military advice from the German Ambassador and government. In the first week of June he ordered the withdrawal of all German troops from Palestine to bolster their efforts in Caucasia. Von Sanders went into convulsions of protest:

> This step will shortly be followed by a collapse of the Palestine front and the political consequences will be Germany's responsibility. With the removal of these bases the Arabs will no longer resist British influence and gold. The Turkish troops here cannot hold the front by themselves, anyone knowing the Turkish army knows they must fail.[251]

Ignored but not finished, however, von Sanders played a last card. To the German Ambassador in Constantinople, he telegrammed:

> Should the German troops be withdrawn from here against my protests, I shall at once lay down the

---

[249] Von Sanders, p. 238. Most of these replacements were new conscripts with no training.

[250] Von Sanders,, p. 243.

[251] Von Sanders, p.243

command of the Army, in order that the name of a Prussian General may not be made responsible for wholly untenable Turkish conditions.[252]

His final, desperate plead for sanity hit deaf ears. He resigned. The Kaiser refused his resignation. He stayed. Some German troops stayed, including a group of storm troopers, but not enough. Von Sanders set up his HQ in Nazareth. From here he exerted command over the Turkish 7th and 8th Armies in the west and the 4th Army in the east – all under strength and under resourced. He retained some German formations and aircraft.

In sport, you have to be offensive to score points and win the contest; the same in business. In war, sitting and waiting is to lose. Von Sanders knew his parlous state. To deal a decisive blow against the EEF it had to be a complete surprise. Despite British air superiority and their intelligence services, on 14 July, Brigadier Cox's 1st ALH Brigade got that complete surprise.

A combined Turkish force with around 1,000 German storm troops hit. 'In spite of the superiority of enemy fliers, the concentration and preparation of troops for the attack remained concealed from the [British] enemy'.[253] That was the best news that von Sanders received. The Germans were in the centre to control the forward movement, with Turks on either side, ordered to keep up.

Unfortunately, the German troops were left in the lurch as the Turkish infantry failed completely. Instead of pushing forward, they halted when the German centre slowed waiting for them to catch up. When the enemy opened with artillery, the centre withdrew to the foot of the hill and the wings at once joined in the rearward movement.[254] A combination of swift ALH counters and artillery halted the advance, converting it to a retreat under fire.

---

[252] Von Sanders, p. 244.
[253] Von Sanders, pp. 249-50.
[254] Von Sanders, pp. 250-1.

Von Sanders had been correct in his assessment of the abilities of the Turkish infantry. Nevertheless, his German troops had also met their match. Australians and Indians routed this force in just a few hours, inflicting heavy casualties and taking 550 prisoners.

The total failure of his force was a major disappointment to von Sanders, whose confidence spiralled. 'Their collapse, together with the ceaseless stream of deserters, made the future of Liman's Army look bleak indeed'.[255] The failure of his Germans really got to him.

Chauvel was delighted, as were the Anzac and Indian troops after shattering the trained and well-armed Germans. And although this attack was defeated within the day, its significance was so immense it is compared with a much longer battle in France, at Le Hamel, that turned the tide of the war. Its impact here shattered the Turks. It devastated the German troops that they could be whipped so quickly. Von Sanders' worst fears were realised. If only the Kaiser had let him go home.

Anzac casualties weren't heavy but they came in a rush. However, to their delight and that of their mates, the medical plan had them through their field ambulances, through the CCS and back to hospital in Jerusalem in little time. Even the enemy wounded received proper medical care that for many was the first time. Astounding the medics, the Turks and Germans who had just finished fighting the Anzacs now turned on each other, necessitating a peacekeeping force that must itself have been a shock to the calm medics:

> Working hard all day, evacuating wounded to the CCS, cars and extra men came from 2nd LH Field Ambulance to help. On a day like this the whole ambulance is in constant action, as a team. Ambulances and stretcher-bearers go out and collect the wounded from the front line and bring them back. They are attended straight away, with the medical officers and hospital staff working flat out all of the time, then loaded onto

---

[255] Hill, p. 160.

ambulances and evacuated at once by road to Jerusalem. Speed in an evacuation is a major requirement.

In the afternoon, Turkish and German prisoners needing attention and feeding were brought in. They had to be kept in two separate groups to stop them fighting and abusing each other.[256]

It is easy to see the reason for disputes between Germans and Turks. As Hamilton continued:

These were the first German soldiers we had seen. They were all well turned out in good uniforms and splendid boots. We were also interested to find that they all carried a supply of quinine for use against malaria, as well as water bottles and other kit in their haversacks. In strong contrast, the Turks were a motley crowd, with ragged clothes, rags around their feet instead of boots, and practically no kit. No wonder they were at loggerheads![257]

There was more to this encounter than the destruction of a few hundred enemy troops. It was the last deliberate offensive attempted against the British in Palestine. It was the only occasion in the campaign when German storm troops were used as an attacking force. When they failed it had a significant effect upon the two rival armies. The fact that a force of German infantry nearly one thousand strong had been so decisively beaten by a brigade of light horsemen was not in itself remarkable, although the Aussies drank out on it for a while. They found on this day, as they had before the attack, that the Germans, although superior to the Turks as offensive fighters, were very inferior with rifles. The constant sniping duels between the light horse and the Turks were usually evenly matched affairs, although when they were charged at in fury, the Turks fell to pieces as we've seen. But Australians sniped at by Germans quickly learnt that they could move into the open with relative safety and rely upon their quicker and straighter shooting to beat the enemy.

---

[256] Hamilton, p. 130.

[257] Hamilton, p.130.

Desert Anzacs

The discord in the enemy camps was again highlighted. The captured German officers complained bitterly that the Turks who were to have attacked simultaneously had betrayed them. They spoke frankly and made no attempt to conceal their disgust.

As the prisoners were marched towards Jericho, then Jerusalem, the curiosity and delight of the populace was boundless. The news of the defeat with its mix of captured Turks and Germans travelled quickly on both sides of the Jordan. From Jerusalem to Aleppo, for the Arabs, the Turkish soldiers, and the general population, the destruction of this German force, greatly exaggerated in the telling, did more to shake Turko-German prestige than anything that had happened since the capture of Jerusalem.

General von Sanders seemed a commander in title only. The Turkish Pashas and his own HQ ignored him. He knew the problem and he knew the solution, but nobody listened. Things got worse after that failed attack in July.

Enver Pasha, the ever-juvenile egomaniac, had waged war on the Armenians and Georgians in the northeastern area of Turkey and the Caucasus after the withdrawal of Russia in December 1917. A diplomatic solution could have been found; but no. Enver wanted domination. With little to be gained but the inflation of Enver's ego, on went the possibility of fighting. To fight, troops and armaments had to be built up in case the fight did resume. They were taken from Palestine where the serious fight was still taking place. Enver was still trying to beat up the British around Baghdad: another futile effort, depriving von Sanders of valuable resources. Back in Palestine:

> Of considerable influence on the conduct of the Turkish troops was the departure of many Turkish officers. A decree of Enver promised to officers volunteering for service in the Caucasus advantages of promotion and double pay. The departure of such volunteers was natural as in this army they received no pay and lived in anxiety about their families. It will ever remain unique in military history that promotion and increased pay

were offered for transfer from a severe battlefront to service in a theatre where no fighting was in prospect for a long time.[258]

This was an astounding act by Enver, guaranteed to accelerate defeat in Palestine. On 1 August, von Sanders wrote to the quartermaster of the Army Group:

> The real reason for the impossibility to hold our front for any length of time is that available resources no longer permit Turkey to conduct simultaneous operations in two theatres so concentric as here and the East Caucasus. Many things that should be supplied to this theatre, we lose to the east. Please inform the German HQ and the Prussian War Ministry plainly of my views. I am unwilling to share the responsibility for false measures that continue to be taken in Constantinople. It is my firm conviction that operations in the east will and must fail. All such Turkish projects since the beginning of the war have failed, but I have never been listened to, and shall not be listened to in this case, until too late.[259]

He must have wished even more that the Kaiser had let him go home. This great Prussian general knew his task was doomed and so was his fate. By now, though, Allenby's sole aim was the complete destruction of the Turkish armies and their German supporters. Morale was high for the EEF, low for the Turko-Germans. But the Turks had shown resilient defence on many occasions, if not worrisome attack. Now they had their backs to the wall, which is the very time when every enemy is most dangerous.

Nevertheless, from the lost battle in July, through August and into September things got worse for the Turkish soldiers and their supply lines. That politicians and diplomats refuse to accept the inevitable and continue to send thousands of their young countrymen to die for no valid purpose beyond their own stubborn stupidity is the tragedy that litters the battlefields of history. At this stage it was clear that the war in

---

[258] Von Sanders, p. 254.

[259] Von Sanders, p. 255.

Palestine had been won, and lost. Nothing was to be gained by further conflict. The only possibility was that more families would lose their young men, never to return to their homes and loved ones. If the fight continued there would be more losers in the fields of England, the outback of Australia, the paddocks of New Zealand, the tribal lands of the Arabs, the plains and mountains of Turkey and the cities of Germany.

But fight on was the Turkish agenda.

Allenby's plan revolved around the Turkish supply, transport and communications systems: cut those and it was all over. There was just the Hejaz Railway to bring supplies from Turkey to be distributed to the three Turkish armies. A branch line from Deraa went west, through El Afule, which was also the centre of the telephone and communication distribution system. The towns of Jenin and Beisan (in today's Israel) were major junctions. The road systems, dilapidated as they were, also went through these towns. Capture those towns and the Turks were without supplies and reinforcements. If Allenby had worked this out, yet von Sanders hadn't and continued to rant on about the pashas and their failure to support his requests, this can only lead to questions about the state of mind of this great Prussian general.

Allenby knew that his mounted troops were the key to success. However, the hilly area in the immediate east was less suitable for mounted operations until it opened onto the desert plains. The open western plains were perfect for fast, mobile horsemen. The Arab Army was in the east, making progress along the Hejaz Railway; rail movement between Amman and Medina had stopped. The Ma'an and Medina garrisons were now so isolated that no support could get in and nothing could get out. The Anzacs were covering the east, but that's not where he wanted them.

Von Sanders believed the thrust of attack would come from the east. This belief was based on the previous raids across the Jordan; that the Anzacs were already across the Jordan; and Arab army activities. He had the under-manned remnants of

Crush the Enemy

two armies in the west and was ready to move them east. He had the 4th Army in the east, near the railway.

Allenby's plan for what he intended to be the final battle ensured that the mobility of the DMC would be exploited to the full and that Chauvel and his commanders would have the opportunities that their horsemen had so long been denied. On 1 August, Allenby had confided his plan to Chauvel, Bulfin and Chetwode.

Bulfin, reinforced with five infantry divisions from Chetwode's Corps, was to shatter the Turkish front on the Plain of Sharon in the west. Through this gap Chauvel would spearhead three cavalry divisions with the aim of cutting off the Turkish 7th and 8th Armies. They would block the roads to the north, then advance on the coastal city of Haifa. Chetwode's infantry was to attack northeast. The rail junction at Deraa was beyond the range of cavalry, so the NAA was to cut the lines north, south and west of it a few days before the main offensive was to start. That early foray could be expected to further distract von Sanders' attention from the main thrust in the west.

The plan was modified over the next few weeks. Allenby's later instructions to Chauvel were:

> having smashed his way through the Turkish front at Sharon; strike northeast; position his Corps 40 miles north of the British lines and one hour ride from von Sanders HQ at Nazareth; cut the railways, seize the crossings of the Jordan and block the retreat of the 7th and 8th Armies. On the Jordan, Chaytor's Anzac Mounted Division was to press the Turkish 4th Army.[260]

Chauvel immediately replied, 'I can do it'. That Chauvel's imagination was fired may be deduced from one of his immediate proposals to Allenby – the capture of the enemy Commander-in-Chief, Liman von Sanders.[261]

---
[260] Hill, p. 161.
[261] Hill, p. 162.

Desert Anzacs

At first he was not in favour, but subsequently Allenby must have liked the boldness of Chauvel's thinking and gave him the nod to develop a plan. Allenby impressed the need for speed and surprise on Chauvel, with no distractions, no matter what opportunities presented themselves: as if he needed to be told.

But now the BIG questions. How to get the DMC from east to west undetected? How to position the infantry and hide their intentions? How to get the Arabs to fulfill their role? He'd have to keep German planes on the ground or shoot them out of the sky. He'd need massive deception. Security was beyond vital. When it started, 'Intense speed and complete surprise would be the order of the day. Any delay would give the enemy time to redistribute his forces'.[262]

The importance of air power had exploded from an unknown, two years earlier, to something of vital importance now. Two jobs for airmen. One, find out about the enemy and shoot down their planes. Two, don't let the enemy find out about us and don't get shot down. Easy enough to say, but how to do it?

Enemy airpower consisted of several squadrons of Germans and virtually no Turks. But the Germans had a few problems. First, the Brisfit was far superior in speed, climb, fire power and manoeuvrability to their planes, and had achieved domination from late 1917. Second, into June the problems had mounted, as shown in a German report captured later:

> The shortage of our machines [from being shot down or destroyed on the ground] is soon to be overcome by the arrival of replacements but pilots are scarce, owing to sickness and other causes.[263]

What's more, the German pilots were reluctant to engage the Brisfits in aerial combat with their now inferior planes. With fewer and fewer pilots overall, German activity fell dramatically. Major S. W. Addison, now commanding No. 1 Squadron, reported that

---

[262] Langley and Langley, p. 149.
[263] Cutlack, p. 134.

> during one week in June hostile aeroplanes crossed our lines one hundred times. In the last week in August the number of enemy visitations had dropped to eighteen. During the following three weeks in September it dropped to four. For several vitally important days before the attack no German machines were seen near the line.[264]

This went a long way to preventing the enemy from discovering much about EEF activities.

Incidentally, Addison, promoted from Captain OC 'A' Flight, had taken over from Dickie Williams who'd been promoted Lieutenant Colonel to command 40th Palestine Wing RAF (the title changed from the RFC). The Palestine Wing included five British squadrons and our No. 1 – another outstanding dominion achievement. The pride of the squadron was summed up in an address given by Major General Salmond commanding the RAF in the Middle East when he inspected No. 1 Squadron in July 1918:

> This squadron ranks as one of the best in the RAF. Its interior economy, workshops and discipline are excellent. The turn-out of its mechanical transport, and above all, of its aeroplanes, are models of their kind. The results that have been achieved by the RAF have been, to a very marked degree, due to the fine work of No. 1 Squadron AFC. It is a matter of pride to me to have had this squadron under my command.[265]

To further complement the skills of No. 1 Squadron, the one and only giant Handley-Page bomber arrived at the Ramleh airfield on 29 August and was handed over to this premier squadron. The piloting of this treasured plane was given to the Wing's premier pilot, Captain Ross Smith, and his main mechanic, Sergeant J. M. Bennett. So, what did the boys do to warrant this chest beating? Highlighting German reluctance to engage the Brisfit and our pilots, Lt Clive Conrick relates:

---

[264] Cutlack, p. 133.
[265] Cutlack, p. 146.

Desert Anzacs

> Today Ross Smith and Kirk, with Paul and Billy Weir flew as far north as Jenin to photograph the aerodrome. They attacked five Hun aeroplanes with their engines running just as they were preparing to take off. At once the engines were stopped and the crews fled for cover.[266]

Although certain squadrons had specific roles, some combative and others not, all aircraft were armed and took every opportunity to drop bombs and blast away with their machine guns. Photo reconnaissance to make accurate maps, hand sketches, firing on trains and stations, monitoring troop movements and camp build ups, attacking troops and cavalry concentrations, artillery fire direction, support to EEF and Arab ground forces were other tasks that had to be undertaken. As a couple of illustrations show, many things could be done on one flight:

> A reconnaissance of great value in reports returned by Fysh and McGinness and Walker and Fletcher left hardly a detail unrecorded of the main routes across that difficult terrain. After finishing that road recon they had what they call 'a day in the country'. At Beisan they found a train; they machine-gunned it and the transport park along side the station. Leaving panic and confusion there, they next put a force of 200 cavalry into a mad stampede. They flew on north to Semakh rail station with a small airfield beside it. With bursts of fire they chased several hundred troops about the station and yards. A dump of flares exploded and started a local fire. Men jumped out of a train; horses bolted in all directions. The Australians had no ammunition left for the sailing vessels on the Tiberian lake.[267]

Mayhem and a good day's work. Conrick reported that incident then went on to report his own:

> At 12.30 the hostile aircraft alarm went off. Nunan and I rushed out to our aircraft, climbing all the way to Jenin. There was no 'archie' so we came down to seventeen

---

[266] Conrick, 9 July 1918.
[267] Cutlack, p. 141 and Conrick's entry of 31 July.

Crush the Enemy

> hundred feet. Still no movement down below so we came down to a hundred feet or so, from which height we machine-gunned seven Hun Scouts and one two-seater on the ground, as well as the hangars. There were troops and mechanics dashing all over the place looking all the world like a lot of rabbits searching for a funk hole. I had great fun and used up four hundred rounds of ammunition. I had two sets of tennis before dinner and won both of them, although only just.[268]

Yes, the British and Australian airmen and their ground crews achieved both aims: discovering the enemy and preventing the enemy from discovering them.

With Allenby's plan to attack the coastal western region, the role of the NAA in the east became critical to supplement Chaytor's Anzac Mounted Division. Lawrence was called to Allenby's HQ from his desert paradise, flown there in a Brisfit by Ross Smith. As Emir Feisal's liaison officer, Lawrence had the job to do to ensure the Arabs both turned up and succeeded in their part of the plan. Their role was to maintain pressure on the Hejaz Railway to ensure that no movement could assist the Turkish garrison at Ma'an and to stop those troops moving north. Then they were to move towards Deraa a couple of days before the main assault, capture the railway station, and act as a decoy to indicate that the east would be the main attack route.

Once more, politics would come into the picture. Britain, with a reversal of political heart, wanted to keep France out of Syria after the war and keep the spoils. The Sykes-Picot Agreement didn't allow this. Meetings in Cairo and a new agreement between Arab delegates and British and French officials, known as the Declaration to the Seven, indicated that whatever lands the Arabs had won by the cessation of hostilities they could keep. Damascus was the target. Allenby had been told this by London. He told Lawrence. He told Feisal. He told Hussein. They all told the other Arabs and an incentive to assist the battle plan was cemented. Back to his desert paradise went Lawrence with hope and loyalty to the Arab cause in his heart.

---

[268] Conrick, 27 August 1918.

In the south was the Southern Arab Army (SAA), commanded by Emirs Abdullah and Ali. Their role was to prevent movement by rail south from Ma'an towards Medina and to prevent that large garrison from offering any interference north.

The NAA and the SAA were supported by the British officers of Operation Hedgehog and a special 'X' Flight of the RAF. To assist X Flight were several Australian mechanics and eventually one Brisfit crewed by Lieutenants Murphy and Hawley. These aircraft strafed, bombed and generally created carnage at stations and forts along the southern Hejaz Railway, destroying rolling stock and buildings and causing casualties. Hedgehog officers with tribal forces and armoured cars destroyed bridges, culverts, trains and buildings. The net result was that Arab forces with their British support prevented any worthwhile assistance from the Turkish garrisons south of Amman.

Lawrence and the NAA moved further inland to Azrak (the location of a marvellous, although now, a somewhat fallen down, Roman fort that is still worth a visit, beside a large wetland on the bird migratory route from Europe to Africa). Joined here by demolition parties of Englishmen, Frenchmen and Indians, this mixed collection went west towards the railway to conduct a series of explosive and destructive raids.

The final offensive was now to befall the worn out Turks and Germans. But how could the preparations develop while keeping them secret? What deception could they dump on Johnny Turk? And what could be discovered about the Turks, their intentions and their strengths, wherever they were?

19: Plan, Prepare, Deceive

If you live near a dragon do not leave him out of your plans
– adapted from J.R.R. Tolkien, *The Hobbitt*

Plan

It's approximately 60 miles from the Mediterranean coast to the Jordan River and 300 miles from the attack start point near Haifa, north to Damascus. That strip is in three sections: the coastal plains (left flank); the central plains and hills then, the Jordan Valley both sides of the river; the eastern plains to Damascus (right flank).

Bulfin's 21st Infantry Corps was on the left and Chauvel had to move the DMC behind them, from the right and past the centre, so that the horsemen could charge through after the artillery barrage and the infantry break through. Chetwode's 20th Infantry Corps was to charge up the middle where the mounteds would sweep around and ahead of them after their breakthrough. In front of the 20th and 21st Corps were the Turkish 7th and 8th Armies. Chauvel's DMC was to get around and cut them off, catch them between the 20th and 21st and destroy them.

Chaytor's Anzac Mounted Division was to remain on the right and link with the NAA with Lawrence and Feisal, then sweep north and south to cut off and destroy the 4th Turkish Army as it moved from Amman and the Jordan Valley with those coming north from Ma'an.

The NAA was to head north before D-Day and capture the rail junction at Deraa.

A four pronged attack. Simple enough.

The trick, however, was to get everyone in position by start day, without being seen to have moved 200,000 men. D-Day was 19 September. The 30,000 horsemen of the DMC had to move all the way from the right to the left. The infantry had to get into position without being noticed. And, at the very commencement, Turkish communications had to be cut swiftly and entirely so no warning could get to von Sanders or from one Turkish force to another.

Prepare

Allenby and his senior commanders seem to have believed that 'preparation and planning promotes positive performance' works for everyone, really. The essentials were secrecy, speed and surprise:

> In past battles Chauvel and Chetwode had been troubled by the lack of security, especially the wild talk about forthcoming operations among certain officers in Cairo hotels. Such irresponsibility could undo all the patient work of deception. All this careful work could count for nothing except a price of lives. This time Allenby gripped all from the beginning; no detail was overlooked and no detail was leaked. Chauvel did not confide the date and time of the battle to his divisional commanders until 17th September, two days before the start.[269]

Past practices of allowing local citizens to wander freely throughout British camps was also stopped, though the build up of stores and supplies in those huge quantities could hardly be hidden. So they were spread around the countryside like a bag of spilt lollies, not too far apart, to hide the exact location of the offensive. The build up of road and baggage transport could quickly move the stores from dumps to required locations.

For Allenby it was vital that his officers clearly understood the plan, the deceptions and the secrecy required. His intimidating size, not to forget his rank and station as Commander-in-Chief, would have been enough to get everyone's attention and

---

[269] Hill, p. 165.

compliance:

> Allenby began a programme of visits to divisions when he explained his plans to all commanders down to lieutenant colonel, impressing his own boundless confidence upon them all. Wavell, Chetwode's Chief of Staff, recalled that 'he radiated victory and undoubtedly inspired those who heard him'.[270]

Control of the air was vital. In eight days before the offensive only one German aeroplane was seen on a photo reconnaissance anywhere near the main concentration of troops. It was chased and shot down. Nothing got to the Germans and Turks. Some enemy air activity against the Arabs occurred way east of the Hejaz Railway that was later nullified by the AFC.

Deceive

Terror: life-threatening and ear shattering explosions from 350 artillery guns. One hundred thousand screaming soldiers with rifles and bayonets. Thirty thousand galloping horsemen with swords or lances like harpoons. This is what Allenby had to conceal before it started.

Many, many things had to be put in place to achieve that surprise, total victory and an end to the war here. And so started the deception.

In the east the NAA bought barley-feed from the Beni Sakr tribespeople around Amman and made contracts for the delivery of sheep, many more than would be needed. They marked out landing grounds that would never be used, with Arabs hired to watch them so rumours spread via tribal networks. The quantities were such that a large force could be fed and supplied.

RAF and AFC aircraft cleared the skies of German aircraft so no observations could be made.

Stores and equipment were moved at night to avoid detection by the dust raised, from the right to the left, then hidden. Trucks and baggage animals would return before morning so their

---
[270] Hill, p.165.

number looked the same to spies and observers in daylight. Extra tents were erected in the Jordan Valley to imitate a troop build up. Night fires were increased.

Troops not needed on the right were moved to the left and kept under trees or shelters to avoid detection. The Field Ambulance of the 4th ALH Brigade were among the early movers:

> On 12th August we stood-to all day. Moved out at 8pm and trekked all night, in a fast ride straight through Jerusalem in the dark. Then watered the horses and on to Enab by 4am. We got about two hours sleep, had breakfast and had to turn out straight away and stand in the sun for two hours for a full Divisional parade for General Chauvel. Moved again at 7pm for a short ride to Latrun, which we reached at midnight. Rested on Wednesday till 8pm, when we came through to Ludd by 1am. We pitched camp in an olive orchard.[271]

They'd left all their tents erected and some troops stayed behind to ride horses backwards and forwards to create dust. Around 15,000 dummy horses were built from sticks and canvas to look as if the real ones were still there, and more arriving in a build up. Highly visible infantry battalions marched down the mountains into the valley through the dust in daytime, then returned by truck at night to do it again the next day.

Extra bridges were built across the river Jordan. The horsemen who had moved to the left camped under trees, in the shade, hidden from enemy aircraft; but few, then none came. Here at last, the troops got some relief from the sun and earthworks and patrolling as they waited for D-Day, but there was no relief from the flies, mosquitoes and crawling critters. Some of the early movers got leave and recuperation. Some got swims in the Mediterranean.

They cooked with smokeless fires of hexamine and dried desert tumble plants.

A big horse race meeting was called for the Jordan Valley

---

[271] Hamilton, p.135.

as recreation for the mounted men for 19 September. The locals would be the spectators and spies passed the notice to the Turks; but no one came, they were elsewhere.

The 36th Australian Stationary Hospital had moved forward to Gaza, the furthest forward nurses and extensive medical facilities could come. From the following observation by Warrant Officer Hamilton during his leave and visit to a nursing friend it's easy to see how secrets can be spilt when care is not taken:

> The nursing sisters know better than the men in the front line when something is about to happen. Days and even weeks ahead, in the Stationary Hospitals all down the line, and in the Base Hospitals in Cairo and Port Said, the beds begin to empty. Hospital ships suddenly turn up and all the seriously wounded are dispatched on their way back to Australia. Less serious cases are evacuated to convalescent hospitals, and those nearly fit are sent to rest camps. Then, whole wards lie empty, silently waiting the arrival of broken bodies.[272]

Even the vets had to prepare for casualties so they got to know when something big was to happen, even if not the exact date and place:

> During the week prior to the offensive, the unit was made fully mobile by the evacuation of cases that had not fully recovered, the handing in to Ordnance of surplus stores and the placing in the dump of surplus kit.[273]

By now the Arabs were in readiness at Azrak, 60 miles east of Amman, but they were attacked from the air:

> German bombs and machine guns struck a heavy blow at the nerves of the natives who scattered in panic at the airmen's appearance and were in danger of complete dissolution.[274]

---

[272] Hamilton, pp. 137-8.

[273] Anon., *History of 10th Aus Mobile Vet Section*, AWM M224 MSS467, undated.

[274] Lawrence, p. 615.

Lawrence took off to Allenby's HQ, begging for relief. Given were three aircraft of No. 1 Squadron, with pilots Ross Smith, Peters and Traill but only two observers, Headlam and Lilly, as Lawrence filled the observers' seat in Smith's Brisfit. Lawrence merrily described the next attack by the German planes while the Aussie crews were in his camp:

> It was breakfast time with the smell of sausages in the air. We sat around, very ready: but the watcher on the broken tower yelled 'Aeroplane up', seeing one coming over from Deraa. Our Australians, scrambling wildly in their yet-hot machines, started them in a moment. Ross Smith with his observer leaped into one and climbed like a cat up into the sky. After him went Peters, while Traill [who was without his observer and gunner, having been left out so Lawrence could get a ride home] stood beside the D.H.9 and looked hard at me. I seemed not to understand him.[275]

Traill was not happy that he couldn't get up into the air to join the party and Lawrence was the culprit, worsening the whole affair by not leaping aboard to fire the Lewis gun! 'He was an Australian, of a race delighted in additional risks, not an Arab to whose gallery I must play' continued Lawrence.

The Australians destroyed several of the invaders and, returning to Lawrence's description:

> Ross Smith was back, and gaily jumped out of his machine, swearing that the Arab front was the place. Our sausages were still hot; we ate them, and drank tea, but were hardly at the grapes from Jebel Druse when again the watchman tossed up his cloak and screamed, 'A plane!' This time Traill won the race, Ross Smith second, with Peters disconsolate, in reserve.[276]

Again, Lawrence wouldn't go up.

Supremacy was achieved with the Germans shot down or destroyed on the ground, but the Arabs were still hesitant. A few days later Smith arrived in the Handley-Page bomber, the

---

[275] Lawrence, p. 615.

[276] Lawrence, pp. 639-40.

Plan, Prepare, Deceive

only one in Palestine. Lawrence describes it thus:

> Before dawn, on the Australian aerodrome, stood two Bristols and a D.H.9. On one was Ross Smith, my old pilot, who had been picked out to fly the new Handley-Page, the single machine of its class in Egypt, the apple of Salmond's eye.
>
> Then our car flashed northwards again. Twenty miles short of Ul em Surab we perceived a single Bedawi, running southward all in a flutter, he yelled, 'The biggest aeroplane in the world', before he flapped on into the south, to spread his great news among the tents.
>
> On arrival at Ul em Surab, the Arabs had flocked in admiration of the great big plane exclaiming 'Indeed and at last they have sent us a "Tiyara" [roughly translates to 'female flying thing'], of which these other things are foals'.
>
> Our men drew from her bomb-racks and fuselage a ton of petrol; oil and spare parts for the Bristols; tea and sugar and rations for our men; letters, Reuters telegrams and medicines for us. Then the great machine rose into the early dusk for Ramleh.[277]

Lawrence used to great effect the immense optimism that came over him to inspire the continued allegiance of his ever-wavering tribesmen claiming, 'the dashing work of the Australians brought thousands of new recruits to the green standard [the Arab Revolt]'.

With renewed confidence the NAA swooped on the railway areas around Deraa on 16 September, three days before the main offensive. Von Sanders, convinced this was to be the area of main attack, started to reinforce his 4th Army in the east, including ordering the 15,000-man Medina garrison to move north. Their commander, Fakhri Pasha, ignored that order and stayed (in fact, he didn't even surrender when the armistice was signed and it wasn't until his officers mutinied in January 1919 that he gave in). Von Sanders also moved a battalion from

---

[277] Lawrence, pp. 638-41.

Desert Anzacs

the coastal plain at Haifa to Deraa, weakening his force in the area of Allenby's attack.

One nervous moment arose, however. An Indian Muslim sergeant deserted and fled to the Turks, taking with him the news that the main attack would be on the coastal plains in the west. This message got to von Sanders around the same time as the NAA hit Deraa. Good for Allenby, bad for von Sanders: the German Commander dismissed the Indian as just another British ploy to deceive him.

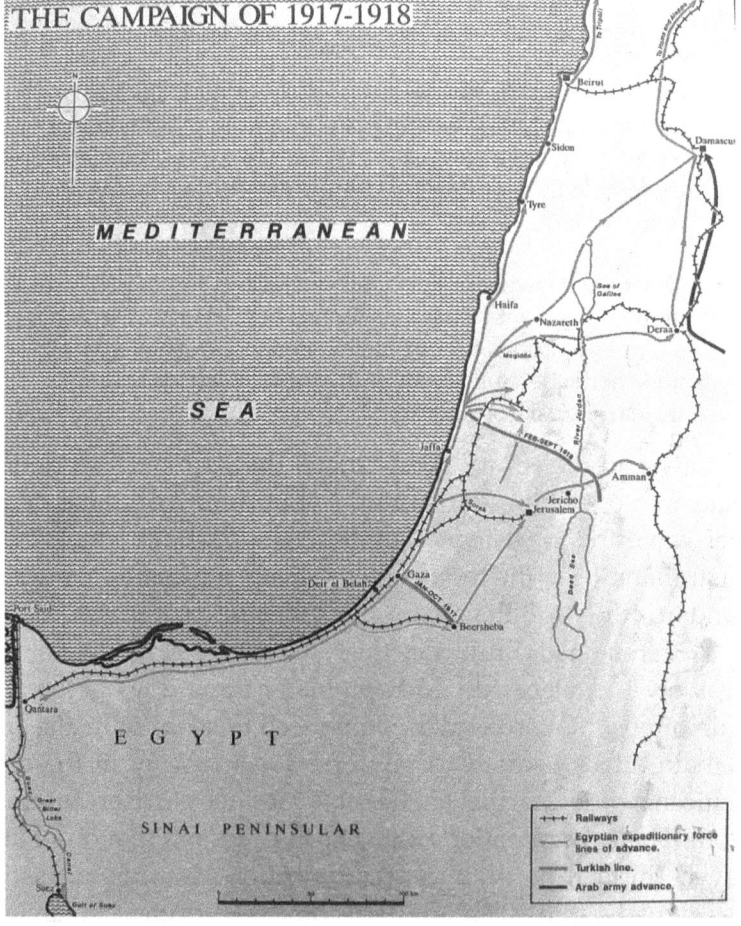

Figure 24: Map of the Campaign of 1917-1918 (from Damascus Commonwealth War Cemetery plaque).

20: Let the Fight Begin

Don't bring a knife to the fight when the other guy has a gun.
– adapted from various sources

The start was set for the early morning on 19 September 1918. The previous day had been hot and the night before was bright with a full moon that fell below the horizon at 3.55am. The breeze fell away as tens of thousands of men groped in the moonless dark, moving into their start positions, soundless in the pre-dawn before the horizon's shutter came up to let in the soft glow of the new day. Tension trapped unspoken words in the dry throats of the infantry. Gunners caressed their tools, soundlessly opening breeches, inserting high explosive shells. Horsemen saddled in silence. Everywhere, nods to mates, no words. Prayers. Thoughts of loved ones. One mile away lay the supposedly unsuspecting Turks. Would this be as swift as they had been led to believe, or would Johnny Turk be waiting ready to cut down the attackers, secrecy vaporised?

It's terrifying to be hit in the dark when sleep is deepest and your reaction is slowest. The advantage of surprise, followed by speed and aggressive action, goes to the attacker. This was Allenby's aim. The greatest reverse shock, however, is to find they are sitting up waiting for you. Who would get the worst shock?

The opening of the offensive is well described, using the words of the time:

> The first blow in the battle was struck by the Australian airmen. Leaving the aerodrome in the Handley-Page at 1am on the 19th, Ross Smith, accompanied by Lieutenants Mulford, Lees and McCann, dropped sixteen hundredweight of bombs on the railway

> junction at El Afule, and during the same night thirty-two hundredweight on the German aerodrome at Jenin. Soon afterwards the Turkish (7th and 8th) army headquarters at Nablus and Tul Keram were heavily bombed, with so much success that the enemy's signal-services were shattered and his communications almost entirely destroyed as the battle was opening.[278]

Moonlight and town lights ensured that the whole of El Afule and the aerodrome were well lit, so takeoff and finding the targets was not difficult. By the time that the bomber had departed for home the telephone exchange was wrecked and the railway station was in ruins. As a result of this raid, the Turkish armies east of the Jordan were out of communications with the Turkish armies west of the Jordan.

Next it was the artillery's task to break open a gateway for the infantry. Gullett continues:

> At 4.30am, while it was still dark, the 350 British guns on the Sharon sector burst into action. The surprise for the enemy, who had been accustomed only to intermittent shelling from seventy pieces, was absolute. For fifteen minutes his trenches were jolted and torn; and his troops, most of whom had been asleep when the storm broke upon them, were in a state of wild disorder when, as the barrage lifted, the four divisions of British and Indian infantry swept towards them with the bayonet.[279]

Then came the mounted force as described by Gullett:

> Immediately in rear of the infantry on the coast was Macandrew with the 5th Cavalry Division; on his right Barrow's 4th Cavalry Division, with the 3rd and 4th Brigades of the Australian Mounted Division, which on the opening day were to be held by Chauvel in reserve, a few miles further back.
>
> To reach the enemy's communications at El Afule, Chauvel's divisions had to ride fifty miles in twenty-four hours; Beisan, on the Jordan, was about eighty

---

[278] Gullett, pp. 692-4.

[279] Gullett, pp. 692-4.

miles from their advanced point of concentration. It was recognised that, provided the infantry made a swift, clean breach in the two systems of trenches, and if Intelligence was correct in its belief that behind those systems the enemy had no reserves of importance south of Nazareth, the mounted divisions had little to do but ride straight and hard on their objectives. Chauvel wisely gave Macandrew and Barrow considerable discretion. Macandrew was to follow the 60th Division, Barrow the 7th. They were only to move after consultation with the two infantry divisional commanders. Once clear of the gap, they were to avoid the struggle and, risking all in their rear, were to speed for El Afule, Nazareth, and Beisan, and rely upon the infantry to drive the enemy back to them.[280]

This offensive became the climax of what most of the men would remember as the worst two and a half years of their lives among the harsh environment and terrain, coping with the local peoples and their cultures, facing the difficulties of food and weather, the enemy and the fighting, the fate of mates left behind or convalesced home, the early difficulties of poor British staff and leadership. Now they had high expectations under good British and Anzac leadership with a definite plan for the end game. Confidence flowed like molten volcanic lava.

A day by day, hill by hill, town by town description isn't warranted here. Not even a battle by battle description: there weren't any of note as the Turks capitulated, drowned by their lack of logistic supplies and evaporated morale. But, there were skirmishes here and there and actions by people that provide the interest, as can the effects.

Such was the speed and weight of the infantry's assault that within four hours the Turkish defence on Sharon was shattered; the wall had been burst asunder for the cavalry to flood through. There was no opposition. The great column streamed northward without even its vanguard being checked. The hearts of all were high as they realised that the plan was unfolding perfectly. While the infantry were fighting their way

---

[280] Gullett, pp. 693-4.

Figure 25: Capt Ross Smith's Handley Page bomber dwarfs two Bristol Fighters, Palestine, 1918 (Australian War Memorial).

through the Turkish defences Chauvel went down to a nearby wadi to bathe; calm, confident and refreshed he then set off in his Rolls Royce in the wake of the 4th Cavalry Division followed by the Australian Mounted Division in reserve.[281]

What has become known as 'the great ride' by the DMC and the greatest cavalry operation of all time had started. This ride should be commemorated as a pillar in Australian and world history by a mixed nationality force under Australian leadership.

The 4th and 5th cavalry divisions under Barrow and Macandrew:

> Moving at the trot, with scouts working wide, the twelve thousand horsemen streamed northwards up the rolling plain. Over the first mile or two the eager Indians had a little play with their lances upon Turkish fugitives from the trenches. Then, lance and sabre glinting in the morning sunshine, the brigades poured on to the crossings of a fine stream flanked in places by wide marshes. From the time Macandrew and Barrow cleared the trenches until they reached Esdraelon they saw nothing of the enemy. Allenby's information was correct. The Turks, confident as to their front line and

---

[281] Hill, pp. 166-7.

Let the Fight Begin

desperately short of troops, had no reserves.[282]

General von Sanders, deprived of troops, reinforcements and equipment was to suffer from the ignorance of the Turkish pashas and his own War Ministry. He also succumbed to the deception and secrecy brought on by his opponent, General Allenby. But surely von Sanders could respond once he knew all hell had unfolded, not in the east, but in the west. To his undoing, the problem for von Sanders was he just didn't know:

> The success of the airmen in blinding the enemy had destroyed the main telephone exchange and put out of action the exchanges of Army HQ. Eighth Army wireless station was knocked out and a standing patrol of fighters over Jenin airfield prevented aerial reconnaissance by the Germans throughout 19th until the Light Horse captured it next day. The Fourth Army at Es Salt remained ignorant of the swift destruction west of Jordan. The magnitude of the disaster was hidden also from von Sanders until Chauvel's cavalry burst into his HQ soon after dawn on the 20th.[283]

However, not everything went according to plan. Chauvel's intention had been to capture von Sanders as Macandrew's Division captured Nazareth. The 13th Brigade, commanded by Brigadier P. Kelly, had led the 5th Cavalry Division all the way from start to Nazareth where they captured 1,200 prisoners and mounds of documents. Seems they stopped on the way to search villages and form flank guards, instead of scooting hell for leather to surround Nazareth as he'd been instructed – no distractions. As a result, von Sanders in his pyjamas made it into his staff car and down the road, barely ahead of the horsemen. Too bad for Kelly, as Allenby on hearing the report, sacked him. Harsh? No, that's how it goes when you upset the big man. Not all von Sanders staff officers were as lucky as him. Some were caught in their PJs still in bed at the Hotel Germania.

---

[282] Gullett, pp. 693-4.
[283] Gullett, p. 697.

The air report gives a clear idea of the unfolding of the first day:

0600. Maughan and Sutherland in a Brisfit escorted a bombing raid. One of the RAF bombers was archied and the engine cut out. The pilot landed at the largest German aerodrome, Jenin. Maughan and Sutherland landed beside them and as the British crew ran towards them a large party of Bedouin appeared, intent on capturing all four airmen. Sutherland fired a machine gun burst at the Arabs as the airmen were climbing aboard. They returned to their base at Ramleh.

0800. Three Brisfits set out on a reconnaissance of the front. After an hours patrol they returned and reported the Turks to be retiring in disorder and at full gallop. They estimated that there were in the vicinity of two thousand cavalry and five thousand infantry and six hundred horse transports in retreat.

0900. All of our machines are in the air flying along the front and continuously bombing and strafing the retreating Turks. Our ground staff are like a goanna at water, continually occupied with servicing and bombing-up returning aircraft.

1000. The eastern recce found all quiet southwest of Nablus with Jacko apparently unaware of what has happened west of the Jordan.

1400. Tonkin and Climie were hit by ground fire while machine-gunning from a low altitude and were forced to crash land and were taken prisoner.

1630. On the road in Wadi Zeimer there was no escape for the enemy forces, the road was littered with wrecked equipment and dead and wounded men, horses and camels.

1900. The HQ of the 8th Army has been captured with many prisoners taken. The rest are in flight.[284]

---

[284] Conrick summarises the first day for No. 1 Sqn AFC, 19 September 1918.

Let the Fight Begin

Back at ground level, the DMC were now behind the 7th and 8th Armies that were in total disarray and streaming northwards, heading for Turkey, their way blocked. At Jenin:

> Satisfied that the Turks would not show fight, Wilson pushed the 10th Regiment out to watch the Nablus road, and Lieutenant R. Patterson of the machine gun squadron was sent with two guns in support. The night was very dark. Patterson at once came into contact with the head of a Turkish column moving on Jenin from the south. He had only twenty-three men and it was clear from the tramp of feet on the metalled road that the Turks were in strength. Trooper T. B. George suggested that they should try to compel a surrender. Patterson agreed and opened machine gun fire over the heads of the approaching enemy. The Turks and Germans, confined on a narrow track between high hills on either side did not know until the machine guns opened and the bullets whistled overhead, that the British were on the plain ahead. As the head of the column halted in confusion, Patterson ceased fire and shouted to them to surrender. By a lucky chance a German nurse who spoke ready English was marching with the officers at the head of the column and Patterson told her he was supported by a large force. She interpreted Patterson's bluff to the officers and, after a brief discussion, the column of 2,800 troops and four guns surrendered to the twenty-three Australians.

The Turks quickly regretted such a hasty surrender:

> The Turks, as they discovered the weakness of their captors, expressed disgust at their surrender, resentfully objected to being herded together, and made spasmodic attempts to escape. But the light horsemen, freely showing their swords, rode confidently among them, and the night passed without fighting. At daylight, when Wilson was able to assemble his prisoners, he found that upwards of 8,000 enemy troops, including many officers of high rank and a few hundred Germans were in his hands together with five guns, several machine guns, two aeroplanes, a wagon loaded with gold and silver money, and a large quantity of other booty.[285]

---

[285] Gullett, pp. 708-9.

Desert Anzacs

Northern Palestine became a scene of mayhem. Horsemen scattered with the speed of wildlife stampedes while Turks fell like leaves from trees in a winter's storm. With the speed of the advance, the ability to continue was based on the ability to receive resupplies. Under the master guidance of Colonel Stansfield, responsible for supplies and transport within the DMC, water and horse feed were plentiful and supplies were arriving regularly; the brigades were ready for fresh enterprises.

A delightful surprise awaited at Jenin. A few thousand bottles of sweet sparkling German wine were discovered in a large cave beside the aerodrome. A light horse guard was correctly placed upon this treasure; but somehow, for some miles around, there was for a day or two a broad smile upon dusty, unshaven Australian faces.

But one tragedy of human survival is that while men on the ground can take prisoners when a devastated army throws its weapons down and its arms up, an airman can't. The airman has little choice but to continue his task of bombing and strafing, presenting him with the dilemma of ongoing killing lest the survivors engage in their own killing of the airman's countrymen and comrades in the future. This became a major moral struggle for many airmen. Worse was to come.

By the end of Day One there were signs of a total rout. Tens of thousands of prisoners, weapons, guns, money, transports and animals had been captured. God knows how many more had died or were wounded. The 7th and 8th Armies, or what was left of them, were in flight. Resistance was scarce.

The British and Indian infantry, supported by West Indians, Ghurkhas and the Jewish brigades had advanced way beyond the original Turkish lines and were in hot pursuit. They were outpacing the wildest dreams of Allenby. Their casualties were heavy early in the morning, then lighter once the Turks commenced flight.

The DMC rode all day. They crossed gullies, wadis, rivers, creeks, through clouding dust then cascading water, up and down mountains, while all the while surveyed by locals bewildered by the speeding horsemen. The light horsemen stood out as their emu plumes fluttered proudly in their headbands. They advanced 50 to 70 miles to capture ground and prisoners beyond anyone's wildest dreams, prompting Chauvel to write his much loved wife about the exploits of his mounted force:

> I have had a glorious time. We have done a regular Jeb Stuart ride. I wrote you two days ago from Jaffa and am writing now from a hill close to Megiddo overlooking the Plain of Armageddon and from my tent door I can see Nazareth, Mt Tabor, El Afule and Jenin. We have been fighting what I sincerely hope will be the last 'Battle of Armageddon' all day. All this is miles behind the enemy's front lines through which Bulfin and Chetwode made a gap for us. It is the first time in this war that such a 'gap scheme' has come off and I am feeling very pleased with myself. No time to write more darling.[286]

Amid the riding and fighting, the DMC casualties had been light. Those who were wounded were quickly attended using the system of treatment and evacuation initiated by Colonel Downes and his medical staffs and the same was true for the vets and the animal treatments and evacuations.

The infantry troops were subjected to progressive depletion when they had to escort the vast numbers of Turkish and German prisoners. Yes, even the almighty Germans were overcome as quickly as the Turks, caught by the speed and dash of the EEF.

Communication between units and commanders is vital. Wire telephone systems laid across the ground by signallers had previously been used and initially were laid here. But the wire layers couldn't keep up with the pace of the sprinting infantrymen, now keener than a feral cat at a fish head to eradicate the Turks. And the flying horsemen were overtaking streams of would-be escapees:

> The pace was too strong for the use of wire communications. To supplement the wireless, Chauvel

---

[286] Hill, p. 171.

Desert Anzacs

had organised a special force of light horse and yeomanry liaison officers, under Lieutenant Colonel E. M. Williams of the Australian forces, to act as gallopers between the divisions and Corps Headquarters. But so swift was the progress that, before the first day had closed, these officers had ridden their horses to a standstill. By the use of wireless and airmen, however, Chauvel was able to keep in intimate touch with his advanced troops.[287]

The vets were now receiving exhausted horses from the dispatch riders, as well as the exhausted and wounded horses from the mounted brigades, in just the first day – what a breakneck pace this was. Fortunately, part of the preparation was to have fit horses well forwards to replace those falling out of the line for whatever reason. With such planning and preparation the battle was able to continue towards its climax.

As the second day unfolded, Chaytor's Anzacs east and west of the Jordan cleaned up the 4th Army that surrendered in droves. They had taken Amman and Es Salt, finally releasing the local friendly population from Turkish oppression. Now he turned his attention to the 5,000-man garrison at Ma'an. However, at the time he was unaware that von Sanders had ordered the withdrawal of all Turko-German forces towards Deraa and Damascus. Some had climbed aboard trains and made good distance until the Arabs got them. Others were on foot and struggled, chased by tribesman waiting at the peripheries for stragglers and small groups that they butchered and looted.

What became known as the Ziza Incident unfolded. The 5,000 remaining Turks from the Ma'an garrison were retreating on foot towards Amman when it was surrounded by up to 10,000 Beni Sakr near the Hejaz station of Ziza, just south of Amman. Around the same time, Australian aerial reconnaissance located the Turkish force, with orders to direct further bombing strikes on the retreating force.

A note was dropped to the Turkish commander advising him that an Anzac force was approaching immediately to their

---

[287] Gullett, p. 697.

north and they should surrender or be vaporised by bombing. Two squadrons (around 500 men) of the 7th ALH Regiment under Lieutenant Colonel Donald Cameron were dispatched to seek and accept the surrender.

On arrival, Cameron noticed hundreds of tribesmen around the Turkish force, unaware there were thousands more behind the hills. He rode on, oblivious of the Bedu but concerned about being outnumbered some ten to one by the Turks. To his relief, a Turkish officer with a white flag came out, indicating that the Turkish Commander, Colonel Ali Bey, wished to meet to discuss surrender.

Ali Bey indicated his willingness to surrender but was greatly concerned that the small Australian force was too small to protect his men from the murderous tribesmen. Cameron became concerned once he realised the extent of Bedu numbers while the Beni Sakr believed the arrival of the Australians was their signal to loot. The tribesmen gathered, like vultures over a roadkill.

Cameron contacted Major General Chaytor and urgently requested reinforcements. Chaytor dispatched the remainder of the 2nd ALH Brigade and set off himself, arriving at Ziza later that afternoon. As Chaytor was assuring Ali Bey they would not be allowed to fall into Arab hands, Brigadier Ryrie arrived with the remainder of the 2nd ALH Brigade. Nevertheless, the growing force of Beni Sakr still considerably outnumbered the total Anzac force. Ryrie now was concerned the Arabs would be emboldened by their numbers and with night approaching, decided there was only one thing for it; add his forces to the Turkish force and stand together.

With the consent of the Turkish Commander and much to the amusement of his regimental commanders, Ryrie organised the deployment of the Australian troops among the Turkish soldiers, allowing them to keep their weapons.

A thoughtful man, Ryrie invited two of the sheiks to come into the camp for a bit of a chat so he could let them know what was going on. Eagerly, they entered. It was then that Ryrie insisted they enjoy his overnight hospitality and informed the two sheiks something to the effect that they would be knocking on the door of Paradise before sunrise if their Arabs attacked.

With this, the Australians introduced themselves to their new trench mates. All enmities were forgiven and they boiled their quarts, made chapattis, swapped smokes and showed photos around the same fires, while Ryrie and Cameron entertained the sheiks.

Early next morning the Aus-Turko force was delighted to see the arrival of the NZ Mounted Rifle Brigade and the departure of the Beni Sakr – followed by the sheiks, now breathing more easily and fresh from their new found friends from Australia and New Zealand. Ali Bey completed the surrender.

The Anzacs captured 4,602 prisoners, fourteen field guns, 35 machine guns, 25 trucks and three trains. The 7th ALH Regiment was distinguished by being the only Anzac unit to have fought alongside the Turks. But Ryrie wasn't finished. As the whole force headed towards Amman, he allowed some of the Turks to keep their rifles, just in case the Beni Sakr returned in larger force. Apparently, the sentries at Amman couldn't believe their eyes as the combined armed force arrived in the city.[288]

In more good news, Conrick notes that Tonkin and Climie reported lost the day before, turned up:

> The Turks captured them while they were burning their plane and took them into a prison. When Jacko left they stayed in prison and were rescued by the British cavalry next day. For prisoners of war, probably the shortest time on record.[289]

Chauvel had established himself as Master and Commander. He'd joined them in Egypt. Took them to Gallipoli and brought them back to Egypt. He led them from the first day across Sinai and was still their leader two and a half weary years later, now at the head of the most victorious mounted force in history.

---

[288] Fenton, D., *Standoff at Ziza*, pp. 18-20.
[289] Conrick, 20 September 1918.

Let the Fight Begin

Figure 26: GARP members at Turkish fort at Fassua Ridge, southern Transjordan.

What was the verdict on Chauvel's performance commanding a force so varied in race, nationality, religion, culture, experience and training?

> All through the long advance, which was to be carried up to Aleppo, nearly 400 miles away, his complete calm, his old-fashioned courtesy to all ranks, and the quiet, even almost languid, tone in which his rapid decisions were expressed in orders, gave his staff and his fighting commanders a steadying touch of inestimable value in that sustained test of endurance, where, from first to last, so much of the mounted enterprise was sheer gamble. Every leader felt that the serene and unpretentious little man at Corps Headquarters, who had led the mounted work unbrokenly from the canal, was an organizer of victory.[290]

By the end of the second day, the completeness of planning, secrecy, training and discipline, the men's dedication and bravery and leadership throughout the whole of the EEF showed:

> The road, clear down the valley of Jezreel and Beisan, was gained and the north-western road from Samaria

---
[290] Gullett, p. 698.

> blocked late in the afternoon, by which time the 4th Division had covered eighty miles in thirty-four hours without off-saddling.
>
> The enemy's general headquarters at Nazareth, and all communications of the Seventh and Eighth Turkish Armies by railway or main roads, were in the hands of his cavalry; only the bridge at Damieh and a few fords remained open. Chauvel's casualties had not numbered a score.
>
> Chauvel's bearing during these two momentous days indicated one of his greatest qualities as a leader. A master of organisation, his plans were laid with a degree of thoroughness which was shown at every stage as the operation smoothly unfolded. Then, satisfied with his work, he awaited the issue with a remarkable absence of anxiety.[291]

It is worth repeating, the DMC casualties were fewer then twenty! Eighty miles, tens of thousands of prisoners, fights and skirmishes. Fewer than twenty is astounding, even given the deplorable state and lack of will power of the Turks, unsupported by their leadership.

It wasn't over yet.

Chauvel was the mounted and flying force commander. He called most of the shots as the pursuit developed, since his force was making the pace and was chased by the support.

Allenby had brilliantly coordinated Chauvel's mounteds with the infantry, maximising his air power that so dominated after the German planes were eliminated, used his artillery at the commencement to destroy Turkish morale that let the infantry race forward unmolested.

With the cavalry encircling just about the entire remainder of the Turkish army in the centre, and the enemy overthrown on the western sector, the task of the 20th and 21st Corps was relatively simple. They were fighting a stricken enemy that was doomed to collapse from its lack of ammunition and supplies.

---

[291] Gullett, p.698.

Let the Fight Begin

Figure 27: Dr John Scott, field archaeologist, viewing the Turkish memorial in Es Salt.

Their morale was lower than the Dead Sea, their enthusiasm lower. Run or quit. Those who quit were escorted to POW cages, food and medical treatment. Those who ran suffered more.

The airmen found columns of men in thousands, horses and vehicles stampeding along wadis, along roads narrowed by the high cliffs flanking both sides. Some fired at the airmen. Others waved white flags. The airmen used their armaments on the leading and rear sections to block the wadis and stop movement. If the ground forces were nearby they'd drop notes and have the surrendering force picked up. But in the case of Wadi Fara, there were no forces nearby and an all-day attack by the RAF and AFC airmen, repeatedly returning for more fuel and armaments, led to more devastation:

> By now Jacko is totally demoralised and has had a gutful of the fight as whenever we came near them they waved the white flags of surrender. It was quite impossible for us to accept the surrender so we just kept on destroying them.
>
> While I kept on firing my guns I had to close off my mind to all that I could see, to the abject terror on the

faces of the Turks, to the dead piling up on the road, the horses falling over the cliffs.

> I love horses but these are the enemy, these are vermin that have to be destroyed before we can have peace in this country. I have the same sort of feeling for the Brumbies on the plains of home, wild horses which I have hunted since I was a small boy, vermin which must be destroyed before they turn that part of Australia into a desert of drifting sands like some of the country I have seen in the Middle East.[292]

But the mood of the airmen changed as the slaughter increased:

> As the entrance and exit of the canyon became blocked with wreckage and bodies, those in the centre of the column floundered, utterly trapped. Australian airmen rained terror on the helpless men for three days, bombing and strafing with impunity.[293]

Nights in the Officers Mess were usually jolly occasions. After this episode it was

> a gloomy night in the mess. Gone our excitement of a few days previously. Gone the elation of having Jacko just where we want him ... We were weary of slaughter ... We were not going to fall down on our job. But oh, those killings! ... Thank God for a bath. That helped – it seemed to wash invisible blood off our hands. Only the lucky ones slept that night.[294]

It still wasn't over. However, the pace of the advance slowed as exhaustion set in and supplies had to catch up lest the infantry and cavalry run out of food, ammunition and replacements. Allowance to evacuate the prisoners, set up civil administration in the captured towns and communications while ensuring units advanced together so none became isolated and an easy prey for a stubborn defender was needed.

---

[292] Conrick, 21 September 1918.
[293] Conrick, 21 September 1918
[294] Sutherland, L., *Aces and Kings*, p. 252.

Let the Fight Begin

Nevertheless, the bold actions of the Anzacs had overwhelmed forces many times their number on many occasions. The 3rd and 4th Light Horse brigades had been left out of the initial attack. When Wilson led the 3rd towards Jenin, the men were elated in their desire for action and their keenness was sharpened by their eagerness with their new swords. They had ridden 50 miles in less than 24 hours and had spent two days and nights with little or no sleep, but they forgot their weariness as they trotted and cantered towards Jenin in the late afternoon. Keeping pace with the light horsemen, the artillery of the Notts Battery jangled along in support.

Major Olden, who was in temporary command of the 10th Regiment, led the advance guard. Shortly before sunset they saw a large Turkish force encamped in olive groves. The Australians instantly decided on their course of action. Lieutenant P. Doig, who had a troop of less than 50 men on the right flank of the vanguard, led his men with drawn swords at the gallop into the enemy's camp. The surprise was complete. The Turks, flustered by the charge, surrendered without a fight. Supported by the rest of his squadron, Doig rounded up 1,800 enemy troops in a few minutes, which included a force of Germans with 400 horses and mules.

Wilson pushed on for Jenin, sweeping round the town and closing the roads leading out north and east. In the twilight they advanced on the town. Bewildered at the unexpected rush, and already exhausted and dejected, and dreading the swords, the Turks everywhere surrendered. As the Australians rode into the streets, however, they were fired upon with rifles and machine guns by men concealed in houses. Throughout the long retreat the Germans had fought bravely in the rearguard of the broken army. This machine gun party at Jenin for a time held up the light horsemen and then endeavoured to escape in the darkness; but after some complicated fighting they were caught by fire from Captain G. H. Bryant's machine gun squadron and surrendered. A few hundred of Wilson's horsemen were now in the dark in the midst of some 3,000 Turks, most of them armed; but the Australians, acting as if they were the whole army, proceeded quickly and confidently with collecting their prisoners.

To the victor go the spoils. Given the speed of the advance, the diet consisted of impromptu 'grab it when you can' meals of bully beef, tinned veggies and tea. But now the villagers produced their harvests of French beans, pumpkins, carrots, turnips, eggs and tomatoes, and even sheep.

At the base of Lake Tiberias was the village of Semakh with its fishing boats, jetties and German machine-gunners to bolster the Turks still there. Hodgson's 4th Cavalry Division was to take Semakh. He ordered Brigadier Grant and the 4th ALH Brigade to take it on 25 September:

> Grant marched by moonlight. They were approaching Semakh just before dawn when they were fired on heavily by rifles and machine guns. 'Form line and charge the guns' was the order. The 11th Regiment drew swords and went for the machine gun flashes at the gallop while their own machine gunners poured in fire from the left. So began a savage and bloody fight that lasted for over an hour. When the enemy surrendered, 100 Germans had been killed; many of the 365 prisoners, of whom nearly half were Germans, were wounded. Australian casualties were seventy-eight and almost 100 horses were hit. It was a costly affair but how much more costly it must have been had Grant delayed his advance to gather in more of his brigade and had arrived in daylight.[295]

In the meantime, on the same day Grant had captured Semakh, Chaytor's Anzac Mounted Division had captured Amman and the rearguard of the Turkish 4th Army. The Ma'an garrison had already been taken at Ziza. In two weeks of action Chaytor's Force had taken over 10,300 prisoners. The casualties of his entire force were 139. 'Such figures reflect more than the collapse of Turkish morale; they show the skill with which Chaytor handled his troops in operations where the enemy

---

[295] Gullett, pp. 730-2.

Let the Fight Begin

always had the advantage in ground and observation'.²⁹⁶

By now the DMC held the line Haifa-Nazareth-Tiberias. Two Turkish armies had been destroyed and Chaytor's force and the NAA had wiped out most of the 4ᵗʰ Army, with the rest fleeing towards Deraa and Damascus and beyond to Turkey.

Yet the airmen were still at it. On 30 September:

> Maughan and Weir on their morning recon over Damascus saw about four thousand Turks, the survivors of six thousand who had been bombed and machine gunned by five of our squadron machines yesterday. They were on the bank of a wadi where two thousand of their comrades had perished. When Maughan and Weir came down to machine gun them these troops just sat quite still with their heads bowed, resigned to die, too exhausted to move. Maughan took pity on them and flew back to base without firing a shot.²⁹⁷

Compassion that day saved many families much grief. And yet, Conrick himself saw it through different eyes:

> When we caught up with them on the road to Rayak they sat on the road with their heads bowed, thinking perhaps that we would let them be, but they soon got a move on when our bullets started smashing into them. I know that some of our blokes had had enough of the slaughter, but for me the war wasn't finished, as I knew that until they surrendered, one Turk left alive could mean another Light Horse casualty.²⁹⁸

In his diary a few days later, Conrick recorded, 'we have just learned that the Circassian cavalry have cut the throats of twenty-five of our wounded and a number more have been shot'.²⁹⁹

There is no innocence, just degrees of guilt.

Part of the Turkish 4ᵗʰ Army plodding its way north came upon an Arab village. The elderly, women and children

---

²⁹⁶ Gullett, p. 730.
²⁹⁷ Conrick, 30 September 1918.
²⁹⁸ Conrick, 30 September 1918.
²⁹⁹ Conrick, 30 September 1918.

were slaughtered in ways too horrific to relate. A day later, tribal Bedouin led by the vengeful Auda abu Tay caught up with those Turks. Again, the story is horrific and there were few survivors. Lawrence was somewhere in the vicinity and whether he did or didn't order 'no prisoners' is disputed. Auda was not renowned for taking prisoners. General Barrow's 5th Cavalry Division had to restore order and remove the Arabs from the scene, attracting some disrespect from Lawrence, who expressed the view that it was Arab culture to take revenge. Barrow's culture was less barbaric.

There was little left to do now, just cut off the flight of the remnants of the 4th Army. The advance to Damascus began with the Turks fleeing through and around Damascus – some on the road towards Beirut in the west and others towards Homs and Aleppo in the north. Chauvel ordered Hodgson to cut off those retreat roads. The road west passes through the narrow, steep-sided Barada Gorge through which flowed the Barada River and a railway line, picturesque at any other time, but restricting road movement to a narrow defile.

The Germans led the bedraggled and worn-out Turkish forces attempting to stream through the gorge and escape to Beirut. By 30 September, the Australians had caught up with the fleeing pack and from their dominant position atop the gorge ordered them to stop and surrender. The Germans refused and forced the Turks to continue while stupidly firing from the lowest position towards the heights. Deplorably and unsurprisingly the result was carnage. The road was completely blocked with carcasses, smouldering vehicles, broken guns and transport wagons. Nothing could move through it, including the Australians tasked to prevent other Turks fleeing north. The only open road went through Damascus.

Who held Damascus at war's end was to control Syria and that was the Arab target. It is disputed even today who entered Damascus first – the Australians or the Arabs. The reality, however, is clear. The Arabs were supposed to for political ends. Lawrence and the Arabs said they did. But all

Let the Fight Begin

their posturing mattered not when the victorious and powerful politicians and diplomats divided the conquered lands in the years immediately after the war. So for the sake of accuracy the real story is told.

General Allenby had issued orders that Allied troops were not to enter Damascus before the Arab Army of Prince Feisal, unless operational conditions warranted otherwise. To catch the retreating Turks the 3rd ALH Brigade had to negotiate the now blocked Barada Gorge. The only alternative was to go through Damascus which, with permission, they did around 6.30am on 1 October.

W.T. Massey, the British official war historian reports:

> The Brigade galloped into Damascus. Major Olden then rode on to the Municipal building, and, finding that organised resistance had ceased, gave directions for the preservation of order. Instantly, half of Damascus came into the streets. The people gave the Australians an amazing welcome. They clapped their hands in truly Oriental fashion, threw flowers and branches of trees on the road, and showered gifts of fruit on the victors.[300]

With this, Major Olden entered the municipal building where he asked to see the civil governor. Emir Said moved forwards and through an interpreter said, 'In the name of the Civil Population of Damascus I welcome the British Army'. After accepting the welcome and surrender on behalf of the British Army, Olden declined the invitation to attend celebratory feasts and activities. With that the ALH remounted and with the aid of a guide provided by Emir Said quickly galloped through Damascus in hot pursuit of the retreating Turks. Their intent was neither to capture nor to hold the city of Damascus.

Massey continues that some two hours after the ALH departure, Lawrence and the forward elements of the NAA arrived in Damascus with the goal of holding and then administering Damascus. Had they been first they believed it

---
[300] Massey. W.T., *Allenby's Final Triumph*, p. 253.

would have given legitimacy to Arab aspirations to attain the permanent administration and control of Syria.

It is possible that at the time of his entry into Damascus, Lawrence was unaware of the earlier Australian presence and took the exuberant local cheering as sign of his liberating entry. The reality would, however, have been brought to his attention in short time. Colonel Wavell indicated that the day before, on the 30 September:

> Within the city, Turkish rule ceased during the day; Arab flags were openly hoisted, and the government was assumed by a committee of Arab notables. At dawn next morning, October 1st, the 3rd ALH Brigade, which had now received permission to enter the city, passed through on its way to the Homs road. These were the first British forces to enter Damascus. Soon afterwards Lawrence and his Arabs arrived, closely followed by the leading troops of the 5th Cavalry Division.[301]

On the day, the Australians probably weren't concerned as to who was first in – they simply wanted to catch the remnants of the Turkish 4th Army. Only later would the political enthusiasts find it of interest. That the Arabs eventually lost Syria to the French, until Syrian independence in 1946, was all to do with supposed peace conferences and the determination of the League of Nations, and nought to do with the Australians being there two hours before the Arab Army – so any dispute should be viewed from either the military or the political aspect to recognise the distinction. And the Anzacs moved on.

---

[301] Wavell, pp. 228-30.

21: Their War is Over

The opera ain't over till the fat lady sings.
– Dan Cook

Brigadier Wilson and his 3rd ALH Brigade took off after the retreating Turks up the Homs road. Seventeen miles beyond Damascus, Wilson had outrun his artillery and run out of food and ammunition, with his men and horses near total exhaustion. Wisely, he called off the chase and camped. The Indians of Macandrew's Division, the Armoured Car Patrol and the NAA took up the chase. In their last two days of fighting, Wilson's Brigade had inflicted heavy casualties and captured over 2,000 prisoners including a divisional commander. Their effort was mammoth. 'Since leaving Lake Tiberias on 27 September, they had enjoyed only one whole night's sleep and, on the last day of operations, had ridden and fought for more than forty miles'.[302] No wonder they outran their support and were falling off their horses. Wilson's march through Damascus and final stop was the end for the light horse, their fighting done.

Beyond exhaustion, the light horsemen fell from their horses like leaves in an autumn wind. Hospital was their rapid destination. Malaria and the worldwide Spanish flu had struck.

Allenby and his senior medical officers knew before the final offensive began that the EEF troops that had been exposed to those malarial mozzies had a fourteen-day window before malaria would cast them bedridden. He had to destroy the Turks within those fourteen days or his cause could be lost. The troops had done just that.

---

[302] Hill, p. 182.

The Spanish flu was something nobody saw coming. That killed over 20 million people worldwide. It created four times more deaths within the DMC than battle casualties had done, with Major Oliver Hogue (Trooper Bluegum) one of those struck down.

Chauvel, however, had two immediate concerns. First, the continuation of the advance to Aleppo towards which the Turks had retreated, but, more alarmingly, a sudden surge of sicknesses. Over 1,200 were admitted to hospital with malaria in the week ending 5 October. In the following week, another 3,100 went to hospital with malaria and the Spanish flu. Even his staff officers and medical officers, including Colonel Downes, went down. Chauvel himself, however, apart from toothache and digestive troubles, avoided both malaria and the flu, thank God.

But the Turks hadn't surrendered and Aleppo hadn't been captured. As it happened, the delay of a couple of days while the light horse lay exhausted and the Indians caught up allowed von Sanders to organise his troops around Aleppo under the command of Mustapha Kemal, the hero of Gallipoli and future leader of the Republic of Turkey to be known as Ataturk.

Allenby ordered Chauvel to capture Aleppo, 200 miles north of Damascus, with his cavalry dwindling and before they all fell over and the Turks got clean away. Only the healthy Indians could do this, together with the British and Australian armoured cars and the Arabs. Barrow's division was heavily weakened and had to halt. Macandrew pressed on with the cars and Arabs. Arriving at Aleppo on 25 October, Mustapha Kemal declined Macandrew's invitation to surrender. That night, 1,500 Arabs under Nuri Bey snuck into the city and surprised the Turks, who withdrew from the city and headed for Turkey. Aleppo was Chauvel's.

Macandrew, having secured the last Turkish city before Turkey itself and recognising the strength of Kemal's force, decided to hold the city rather than chase. Australian armoured cars continued the pursuit, firing some of the last shots of the war in the east. There was little left of the Turkish armies.

On 31 October, Turkey signed the Armistice – the war was over in the east. The Ottoman Empire was gone:

> On Nov 1st, the night sky lit up with star shells all over the place and the word came through about the Armistice, and we knew then that the "copper-coloured devils" had thrown it in at last. Within about 11 days the news of the Armistice on the Western Front came to light. The sky was lit up again with flares and guns roared the salute and you could hear the cheers in the different camps, some far in the distance.[303]

Each year thereafter, at the eleventh hour, on the eleventh day of the eleventh month, Armistice Day is commemorated to remember the price paid for (a temporary) peace.

That was not the end for the Anzacs.

Egypt had provided hundreds of thousand of workers and labourers, artillerymen and soldiers. It had provided food and water in vast quantities. Hearing American President Wilson's calls for self-determination of occupied countries, a strong independence movement grew into a major uprising:

> Soon the country was engulfed by violence, with attacks on trams, railway lines, police posts and military installations. Within a week, the British had lost control of Upper Egypt.[304]

Deprived of their home going, Australians were called on to assist the British quell the violence:

> It was no false rumour as we first thought, and we had to mobilise again, we drew rifles and all our gear and drew horses from the Remount Depots which had been spelling for months. Our chaps were wild at being stopped in demobilising and were eager to clean the 'black devils up'. The ordinary Arab was too ignorant to be blamed; it was the students who were wanting 'home rule' and an Egyptian independence.[305]

---

[303] Bygott, R., diary.

[304] Faulkner, N., *Lawrence of Arabia's War*, p. 459.

[305] Bygott, R., diary.

Many light horsemen, full of dissent and mutinous thoughts, were held in Egypt until May 1919, denied going home to their loved ones. Neither the British nor the Egyptians were too popular at keeping our lads for a lousy police action that was essentially a domestic affair. The uprising finally put down, British control remained, and the Anzacs sailed for their land Downunder.

Epilogue

EPILOGUE

All men dream: but not equally. Those who dream by night in the dusty recesses of their minds wake in the day to find that it was vanity: but the dreamers of the day are dangerous men, for they may act their dream with open eyes, to make it possible.

– T.E. Lawrence, *Seven Pillars of Wisdom*

General Allenby had won an amazing victory. Amazing in its planning, secrecy, deception and execution by his now exhausted and disease riddled soldiers. It's hard to think of a victory in history so complete from preparation to conduct, on such a scale.

In three weeks, from a Turko-German force of over 100,000 there were now 75,000 prisoners, many diseased and sick, while up to 20,000 scampered back into Anatolia; captured were more than 360 guns; 800 machine guns; 3,500 transport animals and countless motor lorries, railway wagons and motor cars. From Jaffa, near today's Tel Aviv, to Aleppo 400 miles away, there was scarcely any sign of the Turk. Only scatterings of busted guns and fragments of shattered wood and twisted metal, deemed too worthless for removal by the ever-needing natives, told of their passing.

In what has been described as 'the greatest mounted ride in history' by near 30,000 horsemen, the DMC suffered 198 killed, 438 wounded, but an astounding 11,300 brought down by malaria and the flu within three weeks of the capture of Damascus.

Of course, this hadn't been a three-week campaign. It had lasted three years.

The Anzacs had been evacuated from Gallipoli to Egypt. They were the one constant in the Sinai Palestine Campaign.

Generals Chauvel and Chaytor and their brigade commanders had been the constant leaders displaying outstanding qualities of leadership and integration of their own soldiers with those of many other countries, of race, colour, religion, culture, language, heritage and attitude, along with a huge native work force.

Medics, vets, nurses, rough riders, drivers, mechanics, storemen, ground crew, cooks, signallers, engineers, postal officers, Commonwealth Bank officers, women of the canteen services and Red Cross, dentists and a naval bridging unit had all contributed to the Anzac role. The flyers, horsemen, cameleers and armoured car drivers were at the forefront of the shooting and performed so brilliantly thanks to their supporters. They all integrated the Anzac spirit and legend that lives on today.

Two things are certain. First, General Chauvel's leadership, Anzac horsemen, their walers and their support soldiery at the Battle of Romani saved the Suez Canal. Second, Chauvel and Chaytor's Anzacs made major contributions to the Arab forces and ensured that tribal Arabs maintained their support of Sharif Hussein's revolt. The Anzacs and the EEF had delivered victory in the East and an open Suez Canal that provided the springboard for victory in the West.

This Great War was the beginning of modern war. It was a war for the scientists, architects, engineers, manufacturers, designers, arms makers and moneymakers. It was a war for the innovators, thinkers and those who could readily adapt to new technologies. Many in the military had to develop new tactics and methodologies for new weaponry to go with proven principles. And politicians soon ensured it was a war to perpetuate war.

In the Middle East, the Sinai Palestine Campaign, a supposed sideshow in the big scheme of the Great War, was a cauldron of political promises, hopes, deceit, intransigence and skullduggery from all sides. Arabs argue they were promised

Epilogue

Figure 28: The original British-centric commemorative plaque to the Desert Mounted Corps.

Figure 29: The final plaque as insisted by the DMC Commander, Lt Gen Sir Harry Chauvel.

British and French guarantees for a unified Arab nation for their support against the Ottomans. They believed they were promised that lands they occupied by military action at war's end would be theirs; hence the rush to be first into Damascus. The League of Nations and non-Arab governments denied Arab control and created a French mandate over Syria and Lebanon. British mandate was given over Palestine, Iraq and Transjordan. Palestine was nominated as the Jewish homeland but not at the expulsion of those Muslims or Christians who lived there. Sharif Hussein did not gain universal, or even modest, tribal or Arab support for his role to be Caliph and King of the Arabs so no Arab nation was, or is, likely. Nevertheless, Arabs believe they were betrayed and accept no responsibility for own disunity due to tribal disputes, sectarian beliefs, generational distrust and those with oil wealth over those without. What is for certain is that the Middle East today is a result of post war political posturing and determinations by the victors, and little to do with the military results.

This Great War was also supposed to be the 'war to end all wars' as world populations cried at the devastation, destruction and death of military and civil landscapes. But the politicians and diplomats of the victors ensured retribution and recompense on the vanquished were so harsh, so stupidly harsh, that within twenty years, World War II would erupt and away the world would go once more to even greater devastation.

But those wars are gone. What is still with the world is the continuation of turmoil in the Middle East. The Sinai Palestine Campaign has had a much longer effect than conflict in the theatres of France and Gallipoli or WWII.

Now, for the legend of Lawrence of Arabia. The American journalist and showman Lowell Thomas, having been bored by the stalemate in France during his war correspondent time in France, had ventured to Jerusalem where he first found Lawrence. In August 1919, he opened a show in London that

Epilogue

portrayed Lawrence as the pale Englishman who had donned Arab regalia and single handedly lead a rampaging Arab Army to momentous victories. Never wanting facts to interfere with a good story, Thomas took his show on a world circuit that appealed to audiences in the millions, wanting a hero to celebrate in the wake of the horrors of Europe. Lawrence became their man.

However, by this time, Lawrence himself had returned from the Versailles 'peace conferences' that [1] ensured another world war was soon to follow and [2] deprived the Arabs any chance of nationhood. He was devastated and hid from the world and the Thomas shows; until he wrote his own book, Seven Pillars of Wisdom, published in 1926. He acknowledged that it was 'my memory of events and not a history of the Arab revolt'. It is the only English language writing of this part of the war and was therefore, taken as gospel. To propel his legend further, the film Lawrence of Arabia hit the screens to an admiring world in 1962, igniting the story once more. But Lawrence had died in a motorcycle accident in 1935. Even today, he still rings as the hero and image of war in the sands of a romantic desert for which we thank Peter O'Toole and Omar Sharrif for their cinematic appeal.

Is Lawrence the hero the world would have of him? Perhaps. He made numerous contributions before the war as an archaeologist and mapmaker, during the war and after as Emir Feisal's adviser at peace conferences, no question. Was he a brilliant military commander, leader and instigator of guerilla warfare as some suggest, or simply able to take advantage of normal Arab hit and run tactics? He was one of a team, a contributor to that group of 50 or so brave and tactful Englishmen of Operation Hedgehog who worked with the Arab Armies.

Edmund Allenby remained in Egypt, was promoted to field marshal and granted the title 1st Viscount of Megiddo and Felixstowe and became high commissioner of Egypt from 1919 to 1925 when he returned to London until his death in 1936.

Harry Chauvel went home and became inspector general, the army's most senior position. The first Australian promoted to the rank of general, he retired in 1930. Recalled in WWII he became inspector general of the Volunteer Defence Corps living until 1945, age 80.

Edward Chaytor became commandant of the New Zealand Military Forces until 1924 when he retired and moved to London living until 1939.

Hudson Fysh and Paul McGinness of No. 1 Squadron AFC formed QANTAS.

Dickie Williams became 'father of the Royal Australian Air Force (RAAF)' and Air Marshall in WW II, retiring in 1946 to become director-general of Civil Aviation. He was knighted in 1954, living until 1980.

Lawrence Wackett remained in the RAAF as an aircraft designer and engineer until 1930, resigning to become head of the Commonwealth Aircraft Corporation. He continued to design aircraft during WWII and was knighted in 1954, living until 1982.

Frank McNamara VC remained in the RAAF and became an air vice-marshal in WWII.

Ross Smith died in a plane crash during an air race in 1924.

Banjo Patterson returned to Sydney to continue writing and broadcasting until his death in 1941.

Robert Bygott, Eric Boulton-Woods, Clive Conrick, Robert Ellwood, Charles Yells, and thousands more young men and women returned to their families and a distant Australian life.

Oliver Hogue, Tibby Cotter, Francis Curran, Harry Wickham and many hundreds of young men didn't go home. Families were devastated and mates were burdened by their own memories and images, rarely to be recounted to those who weren't there – too shocking to relate.

The Young Turks fled, each killed within the countries they fled to within a few years. Turkey would rise again. Mustapha Kemal would bring about and lead the Republic of Turkey

Epilogue

as Ataturk from 1923. Turkish records are prohibited access so little is known of their soldiers who did survive and what home treatment they received. It seems fair to assume it would not have been great, other than the joy of returning to their families.

General von Sanders returned to Constantinople to oversee the repatriation of German troops and was arrested by the British in February 1919, then released to return to Germany where he retired. He lived until 1929, aged 74.

Emir Feisal, expelled by the French from Syria, became the first King of Iraq and Emir Abdullah the first Emir of Transjordan then King of Jordan.

The Sinai Palestine campaign was never visited by our politicians or senior military officers and scarcely saw a war correspondent, only fleeting visits by a photographer and artist. This 'sideshow' to the main event in France, scarcely reported at the time, was thought by an Australian home public to be a rest camp by the beaches of the Mediterranean. It wasn't.

Preserving the Suez Canal and the empire, the oil and food bowl of Mesopotamia, it regained the Holy Land after 730 years of closed Muslim domination. Although disputed by some and acknowledged by other Arab officials then and now, it gave tutelage to indigenous peoples to gain independence eventually, that still has not achieved Arab unity nor given each country administrative efficiency.

More than anything, it showed that Australians and New Zealanders continued the spirit of mateship, pride and national identity begun at Gallipoli. The ANZAC legend, nationhood of two junior, under-populated countries, new on the world stage, had shown the world they could mix it with developed and tradition based nations. It combined the courage and teamwork of those who served overseas and those who remained on the home front who did support them.

Desert Anzacs

To all who went, some to return and some not, we owe a national debt of gratitude and recognition for what has been a neglected and under-reported war.

LEST WE FORGET

Figure 30: Sharif Nasser bin Nasser views his great-grandfather Emir Zeid's uniform and flag from the Arab Revolt at the Australian War Memorial.

Bibliography

Primary Sources: personal papers, diaries, soldiers' books, interviews

Birbeck, Tpr G., diary, MLMSS, Mitchell Library.
Burgess, L/Cpl J., Personal diary, MLMSS 1596/1, Mitchell Library.
Bygott, Tpr R., With the Diggers in the East, family record.
Conrick, Lt C., *The Flying Capet Men*, diary, self-published.
Duguid, C., *The Desert Trail: With the Light Horse through Sinai to Palestine*.
El-Gidman Mohammed Suleiman, interview with John Scott, the author and a translator from the Jordanian Department of Antiquities.
Gerard, Tpr E., *12th ALH Regt, Road to Palestine*, MLMSS 1196, Mitchell Library.
Hamilton, P., *Riders of Destiny: 4th ALH Field Ambulance 1917-18*.
Muaffaq Hazza, (Jordanian archaeologist from the Almasaeed tribe), interview by author 28 July 2015.
Sheik Salami Saleem, interview with John Scott, the author and a translator from the Jordanian Department of Antiquities, 4 April 2011.
Weir, Capt F., diary, MLMSS 1024, Mitchell Library.
Zakariya bin Badhann, (Jordanian archaeologist and historian), interview by author, 22 November 2015.

Primary Sources: Official Studies, Surveys, Newspapers

Commonwealth Bank of Australia, *A Brief History Of Its Establishment*, Chapter XII, Shipping Gold In Wartime, C.C. Faulkner, Sydney, 1923.
Anglo-American Committee of Enquiry, *A Survey of Palestine, Vol I and Vol II*; 1945-6.
Stewart, D., Fitzgerald, J., and Pickard, A., *The Great War: Sources and Evidence*, 2nd Edition, Nelson Publishing, Sydney, 1995.

Secondary Sources: Official Histories, Unit Histories

MacMunn, G., and Falls, C., *Military Operations Egypt & Palestine: From The Outbreak of War With Germany to June 1917*.
MacMunn, G., and Falls, C., *Military Operations Egypt and Palestine: From June 1917 to the End of the War*, Parts I and II.

Official History of Australia in the War of 1914-1918

The Australian War Memorial, *The Official History of Australia in the War 1914-1918*:
- Gullett, H.S., Vol VII, *Sinai Palestine*.
- Cutlack, F., Vol VIII, *The Australian Flying Corps*.
- Scott, E., Vol XI, *Australia During the War*.
- Downes, R.M., *Australian Army Medical Service 1914-1918, Vol I*, Part II, Palestine.

Badcock, Lt-Col G.E., A History of the Transport Services of the Egyptian Expeditionary Force 1916-1917-1918, Hughes Reed Ltd, London, 1925.

Unit Histories

3rd ALH Brigade

5th ALH Regiment

7th ALH Regiment

10th ALH Regiment

6th Mobile Veterinary Section

9th Mobile Veterinary Section

10th Mobile Veterinary Section

Secondary Sources: publications, books, magazines

Aaronsohn, A., *With the Turks in Palestine*, Bibliobazaar, Charleston USA, 2006.

Antonius, G., *The Arab Awakening*, G.P. Putnam, USA, 1946.

Asher, M., Lawrence: *The Uncrowned King of Arabia*, Penguin, London, 1998.

Barr, J., *Setting the Desert on Fire*, Bloomsbury Publishers, London, 2006.

Bennett, W., *Don't Die with the Music in You*, ABC Books, Sydney, 2002.

Black, Tpr D., *Red Dust*, Leonaur, 2008.

Brugger, S., *Australians and Egypt 1916-1919*, Melbourne University Press, 1980.

Dearberg, N., Aussies on Horseback, *Military History Monthly* (UK), Issue 36, Sep 2013.

Dearberg, N., The Arab Revolt and the Anzacs. Journal of the T.E. Lawrence Society (UK), Vol XIX, (2009/10), No. 2.

Dearberg, N., Canvas, Wood, Wires and Tyres: The Story of No.

Bibliography

1 Sqn, AFC in Palestine 1916-1918, *Sabretache, Journal of the Military History Society of Australia*, Vol LV, Jun 2014.

Dearberg, N., The Imperial Camel Corps, *Over The Top* (USA), Vol 6, No.5, May 2012.

Dinning, Capt H., *Nile to Aleppo*, reprinted by Naval & Military Press UK, undated.

Dixon, Prof N., *On the Psychology of Military Incompetence*, Pimlico Press, London, 1976.

Dolev, E., *Allenby's Military Medicine*, Taurus, New York, 2007.

El-Edroos, Brig S., *The Hashemite Arab Army 1908-79*, The Publishing Committee, Amman, 1980.

Erickson, E., *Ordered to Die*, Greenwood Press, London, 2001.

Faulkner, N., *Lawrence of Arabia's War*, Yale University Press, New Haven and London, 2016.

Gilbert, Dr. G.. HMS Suva, Captain W.H.D. Boyle and the Red Sea Patrol, Australian Maritime Issues 2007, Conference Paper.

Grey, J., *The Centenary History of Australia and the Great War, The War With The Ottoman Empire*, Oxford University Press, Melbourne, 2015.

Grainger, J.D., *The Battle for Palestine 1917*, The Boydell Press, Woodbridge, 2006.

Hanioglu, S., *A Brief History of the Late Ottoman Empire*, Princeton University Press, Princeton, 2000.

Hill, A. J., *Chauvel of the Light Horse*, Melbourne University Press, Melbourne, 1978.

Hogue, Major O., (aka Trooper Bluegum), *The Cameliers*, Leonaur, 2008.

Horsfield, J., *Rainbow, The Story of Rania MacPhillamy*, Ginninderra Press, Canberra, 2007.

Hughes, M., *Allenby and British Strategy in the Middle East 1917-1919*, Frank Cass, London, 1999.

Isaacs, K., *Military Aircraft of Australia 1909-1918*, Australian War Memorial, Canberra, 1971.

Johnson-Allen, J., *T.E. Lawrence and the Red Sea Patrol*, Pen and Sword, Barnsley, 2015.

Kayali, H., *Arabs and Young Turks*, University of California Press, Berkley, 1997.

Kedourie, E., *England and the Middle East, The Destruction of the Ottoman Empire 1914-1921*, Mansell Publishing, London, 1987.

Keogh, Colonel E. G., *Suez to Aleppo, Directorate of Military Training (Aus)*, Melbourne, 1954.

Langley, Lt-Col G., *Sand, Sweat and Camels*, Seal Books, Sydney, 1995.

Lawrence, T. E., *Seven Pillars of Wisdom*, Penguin, London, reprinted 2000.

Mallett, R., The Interplay Between Technology, Tactics and Organisation in the First AIF, unpublished MA (Hons) thesis, Australian Defence Force Academy, University of New South Wales, 1999.

Massey, W. T., *Allenby's Final Triumph*, Constable and Co., London, 1920.

Massey, W.T., *The Desert Campaigns*, Putnam's Sons, London, 1918.

Mohs, P., *British Military Intelligence and the Arab Revolt*, Routledge, New York, 2008.

Molkentein, M., *Fire In The Sky*, Allen & Unwin, Sydney, 2010.

Murphy, D., Lt-Col Pierce Joyce and the Arab Revolt, *Journal of the T.E. Lawrence Society*, XXII:1.

Napier, Miss P., *A Late Beginner*, Clockwise Paperbacks, 1966.

Nicholson, J., *The Hejaz Railway*, Stacey International, London, 2005.

Ochsenwald, William, *The Hejaz Railroad*, University Press of Virginia, Charlottesville, 1980.

Parsonson, I., *Vets at War*, Australian Military History Publications, Sydney, 2008.

Powles, Lt-Col C. G., *The New Zealanders in Sinai and Palestine*, Anthony Rowe, Eastbourne, 1922.

Pratt, E., *The Rise of Rail Power in War and Conquest*, King & Son, Westminster, 1915.

Preston, Lt-Col R. M. P., *The Desert Mounted Corps*, Houghton Mifflin, London, 1921.

Rolls, Private S. C., *Steel Chariots in the Desert*, Leonaur Press, 2005.

Schaedel, C., *The Australian Flying Corps 1914-1919*, Kookaburra Publications, Dandenong, 1972.

Scott, J., Conflict Archaeology in Southern Jordan, Wadi Yutm and the Arab Revolt 1916-1918, unpublished PhD dissertation, University of Bristol, 2015.

Scoville, S., British Logistical Support to the Hashemites of Hejaz 1916-1918, unpublished PhD dissertation, University of California, 1982.

Seaward, D., *Wings Over the Desert*, Haynes Publishing, Yeovil UK, 2009.

Bibliography

Sheffy, Y., *British Military Intelligence in the Palestine Campaign 1914-18*, Frank Cass, Portland, 1998.

Storrs, Sir R., *Memoirs of Sir Ronald Storrs*, Putnam & Sons, London, 1937.

Sutherland, L. W., *Aces and King*, John Hamilton, London, undated. Complied from official sources, *The Advance of the Egyptian Expeditionary Force*, 2nd Edition, His Majesty's Stationery Office London, 1919.

Tyquin, M., *Forgotten Men: The Australian Army Veterinary Corps*, Big Sky Publishing, Newport, 2011.

Underwood, Colonel J., Raid on Jifjafa, Sabretache, *Journal of the Military History Society of Australia*, Vol XLII – Sep 2001.

Van Dyk, R., *Sergeant Yells and his work with Lawrence*, Australian War Memorial, Canberra, 2007.

Von Claueswitz, C., *On War*, Wordsworth Classics of World Literature, Hertfordshire, 1997.

Von Sanders, L., *Five Years in Turkey*, Naval and Military Press, London, originally published 1919.

Wavell, Colonel A., *The Palestine Campaign*, 2nd Edition, Constable and Co., London, 1929.

Westrate, B., *The Arab Bureau*, Pennsylvania State University Press, USA, 1992.

Williams, H. R., *The Gallant Company*, Angus & Robertson, Sydney, 1933.

Woodfin, E., *Camp and Combat on the Sinai and Palestine Front*, Palgrave Macmillan, London and New York, 2012.

Secondary Sources: websites, online

http://www.airwar1.org.uk
http://www.anzacday.org.au/justsoldiers/WO2 Noel Bolton-Woodshttp://www.diggerhistory.com.au
http://www.lighthorse.org.au/personal-historieswww.turkeyswar.com
https://en.wikipedia.org/wiki/Charles_Holland_Duell
www.jcu.edu.au/ellwood (removed)
http://www.nzmr.org
http://www.rogerstudy.co.uk/hejaz/tel-airforce/pg1.html
http://www.southsearepublic.org/2004_2002/people/aces/smithross

A great read and need to re-read more than once to absorb the great detail. It's a fantastic gutsy, concise, factual story of the Australian involvement in this unheralded part of WW1. The writings tell in a very acceptable way the detail, without pulling punches of this theatre.

– *Bary Cheales OA, Former Olympian and Retired Business Owner*

"To be totally honest this is a subject I usually go out of my way to avoid, I hate war. All things considered though I did enjoy the read. I really enjoyed the spirit which you captured of the time, the lands and the people involved which I found engaging."

Melissa Morante, Personal Trainer, Newcastle

"I like it. Easy reading. And it's an unknown part of our history that should be known, well done."

Michael Stokes, Human Resource Manager, Thiess Australia

"I found Neil Dearberg's book most informative about a period in Australia's history that should never be forgotten ... Neil has put in an incredible amount of research into this subject and he writes in a way that makes you feel you are actually there."

Fay Chamoun, Director, Floral Art School of Australia and International Floral Design School